European Capital of Culture

CULTURAL POLICY

Edited by Prof. Dr. Wolfgang Schneider

VOLUME 22

Zu Qualitätssicherung und Peer Review der vorliegenden Publikation:

Die Qualität der in dieser Reihe erscheinenden Arbeiten wird vor der Publikation durch den Herausgeber der Reihe geprüft.

Notes on the quality assurance and peer review of this publication:

Prior to publication, the quality of the works published in this series is reviewed by the editor of the series.

Kristina Marion Jacobsen

European Capital of Culture

Cultural Policy Conditions
within the EU initiative,
using the examples of RUHR.2010
and Marseille-Provence 2013

PETER LANG

Bibliographic Information published by the Deutsche Nationalbibliothek
The Deutsche Nationalbibliothek lists this publication in the Deutsche
Nationalbibliografie; detailed bibliographic data is available in the
internet at http://dnb.d-nb.de.

Library of Congress Cataloging-in-Publication Data
A CIP catalog record for this book has been applied for
at the Library of Congress.

Zugl.: Hildesheim, Univ., Diss., 2019

Hil 2
ISSN 1611-700X
ISBN 978-3-631-86425-8 (Print)
E-ISBN 978-3-631-86426-5 (E-Book)
E-ISBN 978-3-631-86427-2 (EPUB)
DOI 10.3726/b18858

© Peter Lang GmbH
Internationaler Verlag der Wissenschaften
Berlin 2022
All rights reserved.

Peter Lang – Berlin · Bern · Bruxelles · New York ·
Oxford · Warszawa · Wien

This publication has been peer reviewed.

www.peterlang.com

English editing: Linda van de Vijver

This publication is the dissertation of Kristina Marion Jacobsen. It was published under the title "Die kulturelle Eigenlogik von Städten. Bedingungen einer transformatorischen Kulturpolitik im Rahmen der Initiative 'Kulturhauptstadt Europas' am Beispiel von RUHR.2010 und Marseille-Provence 2013" in the Department of Cultural Policy at the University of Hildesheim. The dissertation procedure concluded with the academic disputation on 25 October 2019. Prof. Dr. Wolfgang Schneider and Prof. Dr. Olaf Schwencke were the academic supervisors of the dissertation.

Table of contents

Preface

Prof. Wolfgang Schneider
The "European Capital of Culture" (ECoC) initiative, launched in 1984, has become more and more of a driving force during the course of its development. During the bidding process, but even more so after the title is awarded, the responsible actors enthuse about a comprehensive transformation that encourages a variety of positive effects. The initiative has now evolved into a far-reaching programme that triggers new culturally driven development strategies in the participating cities.

In recent years, there have been an increasing number of publications on the transformative potential of cultural policy. In this context, transformation is mostly understood as a comprehensive change and realignment of structures. In terms of transformation, cultural policy is about the social accompaniment of processes, the role of artists and cultural mediators, and sustainable concepts of cultural promotion of infrastructure, project management and cultural education. The European Union's cultural policy also attaches special importance to art and culture within the framework of its identity discourse – although it still sees itself as an economically oriented, supranational organisation. The ECoC initiative is nevertheless a piece of the mosaic in the complex web of European interests.

Criteria for transforming cities and society

"The ECoC initiative offers citizens a framework for reflecting on the role of their city in Europe, to confirm it and to deal with their identity on a small scale (in the surrounding area) and on a large scale (Europe) through artistic activities," writes Kristina M. Jacobsen in the introduction to her PhD thesis. Her intention is to verify how the award criteria led to successful activities which transformed the city and urban society in a sustainable way.

Local politics often place great emphasis on the image and marketing of a city, the increase in the number of tourists, new jobs in the cultural and creative industries, as well as other start-up effects, whereas Kristina Jacobsen focuses on the intrinsic value of culture as it is described in the latest EU decision of 2014 on European cultural policy. Using the examples of RUHR.2010 and Marseille-Provence (MP 2013), she asks about the cultural-political dispositions, to what extent the concepts of the two Capitals of Culture contribute to cultural-political transformation processes at the sites and about the perspectives for further development.

The theoretical framework of the study is provided by the UNESCO Declaration on Cultural Policy of 1982 in Mexico City. At the same time, the so-called New Cultural Policy of the old Federal Republic of Germany emerged with the understanding of cultural policy as social policy, as well as the emphasis on cultural governance, which is also applied in the European multi-level system, as the report of the Enquête Commission "Culture in Germany" of the German Bundestag from 2007 showed. In this context, Kristina Jacobsen also discusses the informal and non-institutionalised regulations and formats of governance. The fact that an active role is ascribed to civil society fits in with the idea of the ECoC initiative.

Cultural policy challenges

The instrument of cultural development planning is being upgraded as a framework for a structural and procedural control system. Kristina Jacobsen dedicates herself to transformative cultural policy, listing the cultural policy challenges of demographic change, diversity and pluralisation of society, as well as digitalisation and globalisation. The involvement of culture is particularly emphasised: "The far-reaching transformation of a city, such as Marseille as a cultural tourism location, would not have been possible if it had been attempted by the cultural institutions alone."

With the theory of the intrinsic logic of cities (after Martina Löw), Kristina Jacobsen provides an approach to understanding cities with their social phenomena as a whole. She is concerned with local cultural identities, with socio-cultural specifics, with interdisciplinary and comparative research. She states: "The ECoC initiative can act as an occasion and a means to promote and contextualise urban development projects, as well as to reflect on the individual specificities of the location and their significance for the future of the city."

"Change through Culture – Culture through Change"

The ambitions and reality of both ECoC years are shown, examples of programmes and projects are analysed, reflected upon and, fortunately, also criticised. The motto of RUHR.2010 was "Change through Culture – Culture through Change" and was intended to trigger the desire for the development of an entire region. In terms of cultural governance, this seems to have succeeded. In this context, Kristina Jacobsen mentions the many cooperation projects with partners in the city and the region. *Local Heroes* turned one of the 53 other cities of the region into a Capital of Culture every week for a whole year. The association of 20 art museums and a network of 11 municipal theatres also served decentralised distribution. *An Instrument for Every Child* was a contribution to musical education that is still supported by music schools and primary schools to this day. The *Emscher Art Trail* remains, as does an annual cultural conference and the *European Centre for Creative Economy (ecce)*. Almost 5 million Euros of state funding are permanently available for the sustainability of the cultural policy developments initiated by the ECoC initiative.

The management's summary is "From myth to brand" – and Kristina Jacobsen therefore rightly states: "This formulation definitely resonates with a utilitarian understanding of culture as an instrument to achieve economic goals." The minimal involvement of local artists is criticised; despite a low-price strategy, the so-called non-visitors were not reached to the extent that was suggested. "Last but not least, the promises made by those responsible for the programme with regard to the sustainable improvement of the cultural infrastructure in terms of mobility have not been fulfilled."

Exchange with Mediterranean cultures

The central theme of MP 2013 was exchange with Mediterranean cultures. A museum financed by the French state (*Musée de Civilisations de l'Europe et de la Méditerranée*) became the face of the ECoC year and also the linchpin of an urban transformation that will continue into the next decade. "In addition to the urban development innovations, [...] the art and culture scene has also experienced an upswing since the ECoC year." The former socio-cultural project (*La Friche La Belle de Mai*) has become a cultural centre for artists and gallery owners. However, Kristina Jacobsen reports on a lack of willingness to cooperate on the part of municipal and regional actors, on the gentrification triggered by MP 2013, and, just as with RUHR.2010, on the dissatisfaction of local cultural practitioners regarding the extent of their involvement in the programme. Apparently, however, MP 2013 seems to have succeeded in enabling the participation of the broad population with low threshold offers under the motto "For free and outdoors".

Kristina Jacobsen attests to the will of both cities to pursue a policy of transformation; she impressively demonstrates the interaction of politics and culture and repeatedly refers to the intrinsic logic of the cities. In 2010 and 2013, ECoC had the potential to stimulate developments in urban society, to make artistic specificities visible and to convey a self-portrait of urban culture. She calls it the "urban self-awareness" that shows that the citizens have a more satisfactory relationship with their city. This also reveals that the public cultural mandate must be continually reflected upon and propagated and must be adapted "to the lifeworld conditions of the present".

Intrinsic cultural logic as a concept of cultural policy

"European Capitals of Culture can serve as an experimental area to try out new structuring and participation possibilities in this sense." The intrinsic cultural logic of cities should therefore also be applied as a cultural policy concept in future ECoC discourses, as should cultural governance and looking beyond the city.

Interestingly, Kristina Jacobsen finds few references to the European dimension in Essen and Marseille. Although the programmes deal with sustainability and participation, the third central criterion of ECoC was the least pronounced in practice. Kristina Jacobsen questions this, as well as the widespread passing on of business models to new applicants for the title and the same cultural advisors, whose activities she already problematises at the beginning as a "travelling circus". She rightly points out that there is no functioning office at EU level where applicant cities can, for example, view and study concepts, documents and programmes. The lack of an archive as a cultural memory of the initiative is also evidence of the relatively passive role that Brussels plays in the whole series.

Kristina Jacobsen is a recognised and astute expert on the discourses and practices of cultural policy. She has been on-site and observed the programmes and interviewed the actors, and she has been to many Capitals of Culture and written about them. She is a co-editor of an anthology published shortly before her dissertation ("Transforming Cities: Paradigms and Potentials of Urban Development within the European Capital of Culture" (Hildesheim 2019)), in which other researchers comment on the phenomena of the ECoC initiative.

Kristina Jacobsen has rendered outstanding services to the cultural-political evaluation of a European initiative, beyond the economic self-interests of the organisers. She demonstrates what is possible to develop the city into a cultural space. With her findings, she takes issue with the approach that municipal cultural policy is a voluntary issue, and provides central elements for a comprehensive urban development concept through the contextualisation of cultural governance.

List of abbreviations

AMU	Aix-Marseille-Université
BIAC	Biennale Internationale des Arts du Cirque
BMR	Business Metropole Ruhr
Bobo	Bourgeois Bohémiens (similar to "hipster" or "yuppie")
CCI	Chambre de commerce et d'industrie (Chamber of Commerce and Industry)
CFSP	Common Foreign and Security Policy (of the EU)
COM	European Commission
Council	Council of the European Union (official abbreviation in EU treaty texts), also called Council of Ministers
CSCE	Conference on Security and Cooperation in Europe
DCFTA	Deep and Comprehensive Free Trade Agreement (Euro-Mediterranean free trade area)
ecce	European Centre for Creative Economy
ECoC	European Capital of Culture
ECSC	European Coal and Steel Community
EEC	European Economic Community
EP	European Parliament
EPA	Etablissement Public d'Aménagement
EU	European Union
IBA	Internationale Bauausstellung (International Building Exhibition)
KMK	Kulturministerkonferenz (Conference of Ministers of Education and Cultural Affairs)
KVR	Kommunalverband Ruhr (name of the RVR until 2004)
LF 2018	European Capital of Culture Leeuwarden-Fryslân 2018
LGV Méditerranée	Ligne à grande vitesse méditerranée

MP 2013	European Capital of Culture Marseille-Provence 2013
MuCEM	Musée des civilisations de l'Europe et de la Méditerranée
NGO	non-governmental organisation
NRW	North Rhine-Westphalia
OMC	Open Method of Coordination
PACA	Provence-Alpes-Côte d'Azur (Region in France)
RUHR.2010	European Capital of Culture Essen/Ruhrgebiet 2010
RVR	Regionalverband Ruhr (name before 2004: KVR)
SGP	Stability and Growth Pact (of the EU)
TFEU	Treaty on the Functioning of the European Union, also known as the Treaty of Lisbon (in force since 1.12.2009)
UNCED	United Nations Conference on Environment and Development
UNESCO	United Nations Educational, Scientific and Cultural Organization
WAZ	Westdeutsche Allgemeine Zeitung
WDR	Westdeutscher Rundfunk

"Culture is not an area free of politics. Culture is not the paradise garden of intellectual and artistic elites. Culture is the way people live. It is the substance that politics has to be about."

(Weizsäcker 1987: 22)

1. Introduction

1.1 About the context of and justification for the "European Capital of Culture" initiative

The genesis of the European Union (EU) is based on a misunderstanding. Even today, more than 60 years after it was founded, this has not yet been resolved. The story of the EU is still told as if it emerged as a pure economic union – but this is not the whole truth. It is of course true that the European Economic Community (EEC) was founded on 25 March 1957 with the Treaty of Rome, as the predecessor organisation of the EU. As the name suggests, the EEC was an economically oriented, supranational organisation into which the European Coal and Steel Community (ECSC), the "Coal and Steel Union", founded in 1951, was transferred. This portrayal of the founding phase of the EU also includes the fact that the economic community was primarily politically motivated, and that a common economic policy between the former wartime opponents was initially intended to pacify the countries of Europe and to better control the coal and steel industries that were essential to the war effort – especially in Germany. Therefore, economic coordination was simply a tool to achieve political goals (see Stratenschulte 2018: 1).

This misunderstanding still plays a significant role in the current EU discourse on identity. "If the Euro fails, Europe fails," said Chancellor Angela Merkel in 2000 in the German Bundestag.[1] These and many other similar quotes from political decision-makers show that the European internal market is to this day still understood as the essential "bracket" of the EU. In the decades when the EU actually contributed to economic growth in the member states, this explanation was not questioned any further. But since the financial crisis in 2007, which led to major disagreements in the European Council with regard to the indebtedness and rescue of individual EU member states, it has become clear that the discourse was not only about money, but also about the foundations of identity and cohesion in

---

1 In the government declaration on measures to stabilise the Euro on 19 May 2000: see link 19.

the EU. Even if the severe crises are now over, the European project has by no means stabilised. On the contrary, other fundamental aspects of the EU such as the membership of individual states (the United Kingdom and Brexit) or the collective allocation of refugees who reach the EU are being questioned. The "European added value" – peace, security, prosperity and legal certainty – is apparently no longer sufficient to dispel the "deep-seated discontent that undermines the foundations of the EU" (Kirchner 2019: 9; see also Stratenschulte 2018: 2).

Not only the growing number of opposition parties on the right of the political spectrum, but even some member states' heads of government, such as Viktor Orbán, rail against the EU and blame it for their problems, even though their countries would probably be much worse off overall without the EU. The numerous and increasingly aggressive voices of the opponents of Europe, and their politics focused on national interests, show that what everything really revolves around is not the debt crisis, austerity budgets or rescue funds, but the fundamental values of Europe.

As the turmoil surrounding the Euro, migration and Brexit allows the EU to continue addressing the crises of the present instead of setting a strategic course for its future role in the international political network, its ideational foundation of solidarity has retreated into the background. The conflict over a common European identity has been the cause of this: "In fact, identity is more than the subject of coffee house discussions, it is the prerequisite for solidarity. The European Union is a community based on solidarity. It builds on helping each other, as happens every day within the framework of structural and agricultural policy" (Stratenschulte 2018: 2). Adapting the above quotation from Angela Merkel, we can therefore conclude: If solidarity fails, Europe fails.

But where and how can we start to prevent the European project from being further eroded? There is undoubtedly a need for a greater awareness of the joint overcoming of the continent's bloody history and the related political, economic and social achievements of the EU since the middle of the 20th century. More cultural and political education, more constructive discourse and more reflective discussions are essential in the current times of crisis.

In this regard, it is astonishing that the already limited budget for cultural policy is expected to decrease for the next multiannual financial framework

of the EU (2021–2027) (see Neundlinger 2019: 30). It is extremely doubtful
that the Culture Committee of the European Parliament (EP) will be able
to enforce its October 2018 request for a doubling of the budget, at least
for the cultural funding programme "Creative Europe" and the successor
programme "Europe for the Citizens".[2] Not least because of the with-
drawal of the United Kingdom from the EU, this EP request is unlikely to
be granted.[3] On the one hand, EU cultural policy is given very demanding
tasks – formulated not least in the decisions on the Capital of Culture initiative that have become more and more extensive over the years; on the
other hand, this policy sector is given only rudimentary funding.[4]

According to the much-quoted bon mot of the long-standing EU
Commission President Jacques Delors – "Nevertheless [...] you don't fall
in love with a large internal market",[5] the EU needs instruments other than
those of economic integration in order to guarantee political stability. The
"European Capital of Culture" (ECoC) initiative has the potential to meet
the challenges of today. It has "proven to be the most popular and citizen-
oriented instrument of the European Union" (Schwencke 2015: 131) and,
according to its legal basis, was based from the outset on "a culture which,
in its historical emergence and contemporary development, is character-
ized by having both common elements and a richness born of diversity",[6]
and later modified and expanded to include the goal of "promot[ing] dialogue between European cultures [...] and, in that spirit, to optimise the

---

2   Opinion of the Committee on Culture and Education of the European Parliament
    of 10 October 2018, available at link 1.
3   Calls for an increase in the culture budget have often been made by the EP, for
    example, at the end of the first five-year term of office of the Parliament in 1983
    in the so-called Fanti Report. This report contained the demand, still unfulfilled
    today, to earmark one percent of the EU budget for cultural purposes, for which
    there was no legal basis at that time.
4   Jens Dirksen (in interview): "Compared to other EU funding grants, the financial
    participation of the EU in the European Capital of Culture is really more than
    ridiculous."
5   See speech by Jacques Delors to the European Parliament on 17 January 1989,
    available at link 4.
6   See Resolution of the meeting of ministers responsible for Cultural Affairs within
    the Council of 13 June 1985 concerning the annual designation of a "European
    City of Culture", available at link 2.

opening up to, and understanding of others, which are fundamental cultural values".[7] The ECoC initiative offers citizens a framework for reflecting on the role of their city in Europe, to confirm it, and to deal with their identity on a small scale (in the surrounding area) and on a large scale (in Europe) through artistic activities.

1.2 The genesis of the European Capitals of Culture

If today we talk about the ECoC initiative as a "saviour" during a difficult period of the EU, there are parallels to the time when the initiative started: "There is no doubt that the initiators intended to use the Capital of Culture idea to strengthen the only vaguely discernible European identity and the support for the European integration" (Mittag 2008: 66). In fact, the idea of the Capitals of Culture was born during the period of the not very successful Greek Council Presidency, which culminated in the failed Athens Summit (4–6 December 1983), at which no agreement could even be reached on a joint final declaration. In order to make a positive impression during this period of generally increasing EU scepticism, manifested, for example, in "milk lakes" and "butter mountains", a largely undisputed counterpart had to be found: a joint cultural project.

Contrary to the assertion, spread in countless ECoC publications and discussions, that the Capital of Culture series was the idea of the Greek Minister of Culture, Melina Mercouri, the idea was in fact that of a European Commission official (COM), "and it was this idea that he taught Melina Mercouri in a quiet hour" (Schwencke 2015: 106 f). Motivated to present the failed Greek Council Presidency as a success after all, Melina Mercouri developed the Capital of Culture concept together with the German Minister of Foreign Affairs, Hans-Dietrich Genscher, and the French Minister of Culture, Jacques Lang. The latter brought in the experiences of the Fête de la Musique, which had already been held annually and successfully in France since 1981, and which was seen as a kind of starting point and inspiration for the new cultural initiative – now in a European context. Melina Mercouri disseminated the concept in the EP's

---

7   According to Decision 1419/1999/EC of the European Parliament and of the Council of 25 May 1999, available at link 3.

Committee on Culture, which had been founded only a few years earlier. With its approval, she was finally able to convince the Council of Ministers of Culture and was thus credited with the Capital of Culture idea in the history of European cultural policy.

From a cultural policy perspective, it is interesting to note that the initiative had no constitutional basis in the Treaty of Rome of 1957. Thus, the decision of the Council of Ministers of Culture of 22 November 1984 on the "European City of Culture"[8] programme represents the first common agreement of the European Community in the field of cultural policy.

During the first five years of the initiative, the ECoC programme took place in already established cultural cities (Athens 1985; Florence 1986; Amsterdam 1987; West Berlin 1988 and Paris 1989), with only few or no cultural or cultural policy achievements. The initiative did not become truly successful until the 1990s, with somewhat smaller ("B") cities using the title to develop comprehensive cultural urban regeneration. With the city of Glasgow (ECoC 1990), the character of the initiative changed in terms of content and impact, as the Scottish city for the first time used the ECoC title as an opportunity for the development of a long-term urban development strategy. "Culture was deployed as an additional motor for the rejuvenation of the city. The power of cultural activity as a stimulus for socio-economic innovation became evident" (Carrijn 2019: 28). The approach of Glasgow 1990, to counteract structural transformation in the post-industrial age with concepts of culture-based urban development, influenced the Capitals of Culture that followed, particularly the locations with comparable socio-economic conditions examined in this thesis: the Essen/Ruhr area and Marseille-Provence (hereafter MP 2013).

Until 2014, i.e. for almost three decades, the legal development of the ECoC initiative followed the practical arrangements of the cities bearing the title: "The entire development of the Capital of Culture has not been an EU development, but the EU has nolens volens let itself be dragged along by it" (Volker Hassemer in an interview, quoted in Jacobsen 2009: 55). Although the regulations were amended and expanded in 1999 and 2005,[9] it is only

---

8   From 1985 to 1999, the initiative bore the name "European City of Culture" and was renamed "European Capital of Culture" in 2000.

9   As can be seen from the extent of the decisions: 1985: one page; 2014: 12 pages.

since the 2014 decision (valid for the period from 2020 to 2033) that the candidate cities have been required to meet differentiated and demanding criteria. These include the following six categories: "Cultural strategy", "European dimension", "Cultural and artistic programme", "Capacity to deliver", "Outreach" and "Management". Also new in this latest decision is the concept of "urban development" and the obligation to professionalise the initiative in principle, which is prescribed by monitoring, evaluation and a European selection jury (which should prevent national influence).

1.3 About the motivations of the ECoC stakeholders

To this day, the intentions of the many different players who contribute to the interactions of a Capital of Culture are very different. However, the success of the initiative is based precisely on the fact that the various political levels can benefit from the Capital of Culture and thus promote their own sectors and interests. The EU has laid down rules for the challenging award, to promote cohesion between the member states in the spirit of the *ever closer union*. At the same time, the latest EU decision of 2014 invokes the intrinsic value of culture: the Capitals of Culture are aimed at "protecting and promoting cultural diversity, fostering interculturality and raising awareness of the value of cultural diversity at local, national and international levels", in accordance with the UNESCO Convention on the Protection and Promotion of the Diversity of Cultural Expressions, which entered into force in 2007 (Decision No 445/2014/EU, recital 3, see link 5).

The position that culture should be promoted for its own sake and should not be used as an instrument for other underlying political mandates is held by Wolfgang Schneider, among others: "The power of art lies in its aesthetic complexity – in the fact that it plays with human sensibilities, reflects reality and discusses questions about our social life. To achieve this, art must be given priority in international politics. Cultural policy therefore belongs on the political agenda" (Schneider 2019: 177). Representatives of this understanding of cultural policy criticise the ECoC development for misusing a European vision for political goals that have nothing to do with culture, such as the elimination of unemployment. "Great ideas are reduced to projects" was the philosopher Hatto Fischer's description of the pragmatism of the participating cities in a background discussion with the author (see also chapter 6.3, footnote 20).

With the ECoC title, local politicians are seizing the opportunity to give new impetus to the urban development of their cities. In contrast to the other two motivations mentioned above, the focus here is on the monetary aspects: the aim is to increase the number of tourists, create new jobs in the cultural and creative industries, and ensure other knock-on effects that lead to an improved image of the location and city marketing. The "European dimension" is of secondary importance or is even completely neglected in the ECoC programme (see Jacobsen 2009: 47).

As the particular interests of the stakeholders presented coincided and merged so well from the very beginning of the initiative, the ECoC series has developed into a successful instrument, the "most established and recognised EU cultural initiative" (Patel 2013: 2) or, in the words of former EU Commission President José Manuel Barroso, the "flagship cultural initiative of the European Union, probably the best known and most appreciated by European citizens" (Barroso 2009: 1, see link 7). The synergies that arise from the convergence of interests described above still contribute to the success of the initiative today, as can be observed in other successful European projects, for example, in the area of the Common Foreign and Security Policy (CFSP) (see Jürgen Mittag 2016: in interview).

However, the interests of the parties involved can also overlap. An example of this is the regional networking of the ECoC RUHR.2010, which in its broad design has pushed the European dimension, which is required by the EU regulations, into the background.

1.4 The structure of this study

The two locations – RUHR.2010 and MP 2013 – were selected because they are considered particularly extensive and largely successful or model-like examples in the development history of the ECoC initiative. They are also, as this study attempts to show, particularly suitable examples to answer the following research questions within the framework of the chosen theoretical-methodological research design:

Q1:   What cultural policy dispositions and concepts formed the basis for the two Capitals of Culture?

Q2:   What cultural policy perspectives for the further development of the initiative were derived from this?

Q3:   To what extent did the concepts of the two Capitals of Culture
       contribute to processes of cultural policy transformation at the
       locations?

The present study attempts to relate the research questions to the two
Capital of Culture sites and to pursue them on the basis of three different
theories: Chapter 2 presents the approach of cultural governance in the
European multi-level system and refers to the ECoC initiative. Governance
structures in the EU and the German federal policy and within the ECoC
initiative are shown.

In chapter 3, the theory of culturally shaped transformation or transformative cultural policy is explained. It is linked to the goals of sustainable development and its content is linked to the other two theoretical
approaches of the analysis.

The theory of cultural intrinsic logic completes the theoretical setting
of this work in chapter 4. It is an adaptation of Martina Löw's "Intrinsic
Logic of Cities" and was first applied to Capitals of Culture.

The methodological approach of the analysis, which is explained in
chapter 5, is based on the expert interviews, which were used to collect
data in accordance with the Grounded Theory. The state of research, the
selection of experts and access to the field are also explained.

Chapter 6 analyses the ECoC RUHR.2010 according to the research
design chosen. As in chapter 8, which is devoted to an analysis of the
ECoC MP 2013, the research design is based on the triad of criteria ac-
cording to which a Capital of Culture is to be assessed, according to Olaf
Schwencke: sustainability, participation and European dimension (see
Schwencke 2005: 28–36; Schwencke & Rydzy 2004: 8–9; Jacobsen 2009: 4).

Between the analysis of the RUHR.2010 and MP 2013 sites, chapter 7
explores the importance of rural areas within the ECoC initiative. This
aspect has often been the topic of ECoC conferences in the recent past (see,
inter alia, Schneider & Jacobsen 2019) and shows new fields of action, but
also the capacity limits of Capitals of Culture.

The broad concept of culture used throughout the research corresponds
to the 1982 UNESCO Declaration on Cultural Policy in Mexico City (see
link 29). Furthermore, an understanding of cultural policy as social policy
is the starting point of the study.

2. Cultural governance in the European multi-level system

2.1 Reasons for selecting the governance approach

The "European Capital of Culture" (ECoC) initiative offers a profitable field of application for governance research. As a funding instrument of EU cultural policy, which can only act cautiously in the cultural sector anyway, due to the subsidiarity principle, it provides an open and flexible framework within which participating cities can create an individual governance system. The assessment of Schneider and Gad – "concepts of good governance for cultural policy have received little attention in the past" (Schneider & Gad 2014: 5) – can thus be understood, in relation to the chosen research topic, as a mandate to focus on the innovative governance structures which various Capitals of Culture have established in the recent past and have successfully used to achieve their objectives.

2.2 Governance in EU (cultural) policy

Governance is an English term that has been translated into German scientific language and for which there is no common German equivalent. Its etymological origin is Greek (κυβερνάω: steer, control the rudder) or Latin (*gubernantia*: steering, management, governance) or compare the French: *gouvernance*. In current scientific usage, the ambiguous term "governance" is used to refer to "the totality of collective regulations which are aimed at a specific problem situation or social situation and are justified by reference to the collective interest of the group concerned" (Zürn 2008: 554).

In EU research, the term "governance" has been used since the 1990s to describe particularly non-hierarchical forms of political governance in the European multi-level system (multi-level governance) (see Große Hüttmann & Wehling 2013: 220). But even before then, governance had already become the focus of EU policy-making in all its fields of action. Since its introduction in 1974, and especially after the establishment of the economic and monetary union with the Treaty of Maastricht (1992), the European

Council has repeatedly introduced additional procedures to coordinate nationally based instruments. Both in the "Lisbon Strategy", adopted by the European Council in 2000, and in the Stability and Growth Pact (SGP), which is intended to ensure compliance with fiscal policy discipline, modes of governance are being used (Wessels 2008: 366 f). They exist in different forms, but at the same time share common characteristics. One criterion for distinguishing between the various procedures is the degree to which they are binding. Depending on the degree to which deviating behaviour can be sanctioned, the terms "soft coordination" or "hard coordination" are used.

Policy coordination in the context of multi-level governance is often understood as a "third way", which, with its innovative methods, lies between supranational regulation according to the Community method, on the one hand, and conventional intergovernmental cooperation, on the other.

> Integration-oriented expectations see these efforts as an important stage for further deepening, while sovereign-oriented guiding ideas understand the use of these procedures to ward off further shifts in competencies. Successes or failures thus also influence the debate on the further shaping of the institutional architecture (Wessels 2008: 366).

Beyond the "old institutionalism" based on treaty law (Schneider & Aspinwall 2001: 3), governance in the EU context refers to the interaction of various actors, in particular parties, associations, NGOs, representatives of regions, local authorities and third countries, as well as associations and media involved in EU decision-making processes. Their activities within formal or informal networks are known as "network governance" in the scientific language of European studies (Wessels 2008: 31). The governance concept is based "not on the separation of tasks and responsibilities, but on the division of responsibilities and cooperation, i.e. partnership of responsibilities" (Knoblich & Scheytt 2009: 4).

The conceptual content of governance opens up through its differentiation from traditional forms of governance, i.e. a formal dimension of politics defined by constitution, law and statute as well as institutions of governance endowed with the power to enforce legitimate political decisions (see Nohlen 1998: 236). Governance, on the other hand, also includes informal and non-institutionalised arrangements and forms of governance that "people and institutions either have agreed to or perceive to be in their interest" (Commission on Global Governance 2005). Overall, governance aims to improve the management of a political or social entity in achieving its goals. Thus, governance also includes elements of corporatism, according to which the organisations, units or individual actors engage actively in the management of the respective field of action.

In terms of the governance approach, both an interconnectedness of political decision-making levels and innovative modes of governance are emerging, such as the Open Method of Coordination (OMC) in the EU. It is applied in policy areas where the EU has no real competences so far (such as cultural, health or social policy). The OMC is also referred to as "soft law", as it comprises only intergovernmental agreements at the level of the European Council. It is applied on a voluntary basis of the member states, for which there are only mutual benchmarking procedures without sanctioning mechanisms.

In its 2001 White Paper on "European Governance" (link 15), the COM emphasises the particular importance of the broad participation of civil society in EU governance. Große Hüttmann and Wehling thus regard governance as an unofficial guiding principle of the EU (Große Hüttmann & Wehling 2013: 221). In 2012, the EP also mentioned governance as a principle of action for the cultural and creative sector in an issues paper and recommended the following for the future orientation of the EU cultural funding programme "Creative Europe": "Develop new approaches to sector assessment, governance and budgeting in the context of the new policy and economic environment for culture and creativity."[10]

In the Communication from the Commission "A New European Agenda for Culture" of 22 May 2018, the COM formulates its plans to–

> broaden the current structured dialogue, going beyond topics examined under the Open Method of Coordination, making more of online collaboration opportunities, and opening up to relevant organisations outside cultural and creative sectors on a case-by-case basis. It will also propose a more active role for civil society in preparing the biennial European Cultural Forums.[11]

---

10 European Parliament 2012: "The Culture Strand of the Creative Europe Programme 2014–2020", chapter 4.1.3, s. 42, see link 17.
11 European Commission 2018: "A New European Agenda for Culture" (COM (2018) 267 final), paragraph 6.2, pp. 9–10, see link 14.

This approach, which also explicitly involves civil society in a governance sense, is taken further by the Council, which, in its Work Plan for Culture 2019–2022 of 21 December 2018, also outlines other less formalised means of exchange and consultation in addition to the OMC:

> ad hoc or Commission-led expert groups, peer-learning activities, studies, conferences, stocktaking seminars, the European Culture Forum, dialogue with civil society, pilot projects, joint initiatives with international organisations, Council conclusions and informal meetings of officials from Ministries of Culture and, if appropriate, from other ministries.[12]

This direction of development shows that more participatory forms of decision-making are favoured in EU cultural policy. They suit the policy field, which is to be treated sensitively here due to the subsidiarity principle and not least because of the great emotional involvement of citizens.

The Treaty of Lisbon, which came into force in 2009, reformed the contractual basis for decision-making procedures in the EU. Compared to the previous legal basis of the EU, the fundamental abandonment of the principle of unanimity in the Council became particularly significant, which also had implications for the cultural sector. This is because the majority decisions thus introduced for votes make possible new discourse coalitions between member states, the COM and the EP, which affect the competences of national and regional actors even more than before. In this way, new forms of cultural policy governance of the EU have emerged, which can promote further communitarisation in cultural policy (see Singer 2010: 23).

2.3 Governance as a principle of cultural policy action in Germany

As a national administrative concept, governance belongs to the concept of the *Activating State*, which was developed in the 1990s and replaced the idea of the *Lean State*. To solve social problems, the *Activating State*, according to the governance approach, involves civil society wherever possible and does not see the state as the only responsible entity. In this context, the

---

12 Council Conclusions on the Work Plan for Culture 2019–2022 (2018/C 460/10), Annex I, E III, see link 16.

public administration has the task of ensuring that there is no "atomisation" of the individual institutions (see German Bundestag 2008: 126 f).

In Germany, governance is practised through the federal system in a multitude of network-like structures, including in the field of cultural policy. The policy interdependence between the federal government, the states and the municipalities continues to increase. The final report of the Enquete Commission of the German Bundestag on "Culture in Germany", adopted in 2007, consequently proposes the approach of cultural governance as a basic model of cultural policy (German Bundestag 2008: 128 f). It emphasises the importance of cultural development planning (Kulturentwicklungsplanung) in several places. This is to be expressly advocated as an important instrument of the states (Länder) and municipalities (German Bundestag 2008: 142). It also emphasises the functioning of cultural governance:

> Such cultural development planning is developed by politics in the sense of the governance approach in a broad social discourse with governmental and non-governmental actors. Cultural development planning defines normative goals, sets priorities and regulates responsibilities. It ensures transparency in cultural policy decisions and thus strengthens the steering and control function of the legislature (German Bundestag 2008: 93).

With regard to cultural development planning, which has become one of the most important municipal cultural policy instruments in Germany, governance in its structural and processual dimension of control and regulatory systems proves to be a successful framework in which goals in the field of cultural policy can be achieved. This involves not only the state (the "first sector") and civil society organisations (the "third sector") but also the private sector (the "second sector"). In public–private partnership projects, joint responsibility is assumed for political action, for example, in cultural infrastructures or individual, limited cultural programmes such as the ECoC year. Depending on the situation, the three sectors play a more important or subordinate role.

Wolfgang Schneider points out that cultural governance is therefore not an end in itself, but serves the question of which management and control systems can be achieved through cultural policy: "Per se, governance is neither a goal of Cultural Policy, nor a sphere of activity within Cultural

Policy, but serves to question what systems of management and control are actively involved in the attainment of Cultural Policy" (Schneider 2016: 3).

2.4 Governance as a steering principle within the ECoC initiative

The EU regulations for the ECoC initiative have been specified and expanded several times since the first decision on the initiative in 1985, and up until the most recent legal basis of 2014. Until now, it is entirely up to the cities themselves how the criteria are to be developed, which can be seen, for example, in the responsibilities for governance: Is the management of the ECoC located in the culture department or does it act as a separate entity? Is it bound by instructions from the municipal cultural administration or higher political levels, or is it largely independent? Likewise, in budget questions, the free design possibilities are recognisable: there is neither a minimum budget for participation, nor is it regulated a priori which sides have to contribute which part to the financing. Other coordinates of ECoCs, such as their content-related focus or geographical extent, are also left up to the candidate city.

"Co-governance" at various levels, which is subject to neither the top-down nor the bottom-up principle, takes place in the sense of cultural governance in network structures with centres of power, cliques and peripheries: "Networking is the new principle in the world of culture" (Schneider 2019: 178). When analysing governance structures in the context of the ECoC initiative, the instrument of network analysis can therefore be helpful, but it will not be discussed in detail in this paper.

In all process-immanent tasks (financing, institutional and artistic cooperation, programme organisation, etc.) of ECoC planning and implementation, the establishment and expansion of networks play an important role. Every ECoC requires–

'multi-project management' due to the numerous projects and the cooperation with a large number of participants, all of whom have to be managed or at least coordinated at the same time. The complex cultural management and the interest-driven cooperation with heterogeneous actors in self-organising networks can be characterised as a steering model of cultural or regional governance with the aim of integrative urban development (Baier & Scheytt 2011: 155).

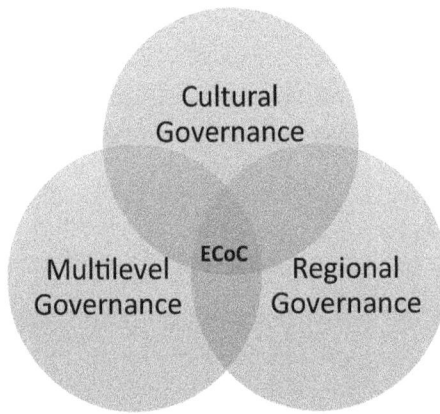

Figure 1: The intertwining of regional, multilevel and cultural governance

Within the ECoC initiative, all three terms – cultural governance, regional governance and multilevel governance – describe, in the intersection of their fields of meaning, a network of actors who aim to develop sustainable cooperation and networks that will promote culturally shaped urban development and stable structural support for art and culture.

In the context of the research on the ECoC initiative, it should be particularly emphasised that the governance of the participating cities and regions has a normative dimension (see chapter 6.4), which depends on ideas of order and intrinsic logic (see chapter 4) that are jointly agreed upon or at least followed, even if they are not based on a formally decided legal foundations. These can be strategic goal formulations or guiding principles, such as cultural development plans, which cover not only individual institutions but, for example, the sphere of action of an ECoC as a whole. They are developed in advance in a broad social discourse with governmental and non-governmental actors and define goals, priorities and responsibilities. This process guarantees the necessary transparency in cultural policy decisions and strengthens the steering and control function of the legislature. Ideally, contracts are then concluded with the institutions and their fulfilment is monitored through reports and control mechanisms. Thus, in addition to the (institutional) framework conditions, governance also includes material specifications and procedural elements such as transparency, reporting obligations and cooperation requirements. If this is applied to a cultural institution, for example, it is then given largely autonomous freedom of action, but must be accountable for its results to the responsible political level. Despite the autonomous actions of individual actors, the governance approach always means that there is a fundamental joint responsibility of politics, administrations and cultural institutions with regard to the subject matter (see above).

3. Transformative cultural policy

3.1 What is transformative cultural policy?

In recent years, there have been an increasing number of publications on the transformative potential of cultural policy (see, among others, Schneider et al. 2019; Sievers et al. 2016). Underlying the reflections is an understanding of transformation that brings about comprehensive changes and realignments of structures. Sociologist Raj Kollmorgen's definition of "transformation" as a "factual and temporal totality of specific and relatively purposeful processes of social change" (Kollmorgen 1996: 283) can also be applied to the cultural sphere.

It has been shown that in addition to the processes of transformation in politics, the economy and society, the cultural sphere is an essential component of change:

> Cultural forms of expression not only reflect and comment on processes of social and political change, they also contribute to change themselves, to an opening of social discourse. Culture and development or transformation seem to stand in a dialectical relationship to each other and are mutually dependent (Schneider 2019: 176).

In general, the following are some of the cultural policy challenges that have occurred in recent years in most cities that require transformative cultural policies:[13]

- demographic change, diversity and pluralisation of society
- digitalisation and globalisation
- changing communication and participation behaviour of culture recipients
- sometimes parallel (over-)supply of similar content
- new, often invisible cultural actors
- lack of network structures
- lack of cultural education.

---

13 See Föhl & Wolfram (2016: 32), modified by the author.

If the "New Cultural Policy" that emerged in the 1970s understood reforms more as an expansion, primarily with a socio-political function (see Heinrichs 1997: 22 ff; Klein 2007: 61 ff), the transformative cultural policy of the present stimulates a comprehensive change that also changes the narrative of the initial cultural policy situation. It is thus a matter of turning away from the hitherto rather additive logic of cultural policy procedures, since unlimited growth in the field of cultural policy is not possible anyway. Instead, it involves changing the existing structures and conceptions in order to both enable the new and protect the tried and tested on this basis. Because the transformative approach affects existing thought patterns as well as the identity and affiliation paradigms of those involved, the process is a priori complex and delicate (see Knoblich 2018: 87 f).

An example of and at the same time a vehement demand for the transformation of cultural policy is provided in the book *Kulturinfarkt* (Haselbach et al. 2012). The authors call for a rejection of the supposedly enlightened attitude of improving people through a culture that is in fact incapacitating and statist. "Cultural sovereignty" is a sovereign act of the state and is authoritarian and out of place in the cultural sphere. The plea of *Kulturinfarkt* ("cultural infarction") certainly does not unjustifiably attest to a one-sided programmatic orientation of the cultural sector; however, the cultural state does not have to be completely abandoned in order to address the frustration of citizens and initiate the transformation of cultural policy (see Knoblich 2018: 88).

3.2 Approaches to transformative cultural policy in ECoCs

In cities that are preparing an ECoC bid or have already been awarded the title and are in the process of preparing the programme year, the above-mentioned general criteria of transformative cultural policy are addressed. In addition, the site-specific deficits and opportunities that make a transformation process necessary are considered. If a transformation is not set in motion by a seemingly independent process such as globalisation, but is intended politically a priori and on a limited territory, the following questions will arise in advance: What is to be changed and with what aim? What is the change needed for? Why is it necessary and who will shape it? (see Föhl & Sievers 2013: 63 ff).

In the participating cities, a variety of municipal fields of action (culture, economy, social affairs, education, tourism and many more) are integrated into the ECoC strategy. Here, the questions about the need for transformation are at the forefront; they must have already been answered in the preparation of a bid book. With the exception of the city of Donostia – San Sebastián (ECoC 2016), whose entire ECOC programme was focused on the theme of peace and which, according to the general director Pablo Berástegui, did not focus on generating creative economic growth, (see Jacobsen 2016: 18) the other cities primarily opted for the programme to trigger a general upswing that went far beyond what they had previously wanted to improve with "cultural promotion".

With regard to the "European dimension" demanded of ECoCs, only the discussion about the transformation of the cultural sphere evokes fears: after all, culture is seen as the foundation stone of Europe, as something that must be preserved and that constitutes the identity of the European *demos*. The cultural heritage of the continent, whose diversity has been legally protected since the Maastricht Treaty (1992) with the so-called "cultural compatibility clause",[14] seems to be at odds with the dynamics and questioning brought about by transformation. The Lisbon Treaty, the EU's most recent legal basis, also invokes the importance of culture for the EU. Its preamble states that it "draw[s] inspiration from the cultural, religious and humanist inheritance of Europe, from which have developed the universal values of the inviolable and inalienable rights of the human person, freedom, democracy, equality and the rule of law" (Treaty of Lisbon, preamble, see link 9). The common cultural values in the EU, which have grown through the inseparable histories of the member states, are evoked daily as a stable basis, especially in relation to political change, compared to the trade policy activities of the governments of the USA and China, for example, which affect the world order. But how can the existing be preserved and the new developed at the same time? Often there is not only a lack of funding, but also a lack of courage to break with tradition and forge new paths. "Such a willingness to transform is an honest, rational, but at the same time brutal

---

14 "The Union shall take cultural aspects into account in its action under other provisions of the Treaties, in particular in order to respect and to promote the diversity of cultures" (TFEU, Art. 167,4, see link 10).

demand. It forces us to see things clearly, to evaluate them and also to question them. It is a 'disenchantment' of cultural politics as a gesture of goodness that must not be questioned" (Knoblich 2018: 88).

In the ECoCs, approaches to culturally shaped transformation are being tried out in the spirit of cultural governance, which presupposes cooperation between different actors from civil society, the public sector and the private sector – not only during implementation, but throughout the entire process from initial planning to final evaluation. The far-reaching transformation of a city such as Marseille as a cultural tourism site would not have been possible if it had been attempted by the cultural institutions alone:

> The reason for this is by no means arbitrariness, but a specific form of cultural complexity, in which it is only through the different understandings of art and culture in different social sub-sectors that a view of the whole and new partnerships become possible or new synergies are created (Föhl & Wolfram 2016: 33).

The example of the ECoC initiative in particular shows how important it is for the cultural transformation process of a city to be the joint task of the various municipal actors.

3.3 Sustainable transformation

Examining the possibilities and results of transformation processes within the ECoC initiative means looking at their sustainability. Sustainability is understood here as sustainable development "that meets the needs of the present without compromising the ability of future generations to meet their own needs" (United Nations 1987).

The debate on sustainability began at the beginning of the 1970s. The report "The Limits to Growth" (Club of Rome 1972) and the beginning of the environmental movement (which brought frightening terms such as "forest dieback", "nuclear fallout", "oil crisis", "species decline" etc. into the social consciousness) represented a turning point in Europe with regard to the previous belief in infinite growth. The discourse on sustainability that began at this time was accompanied by a call for renunciation (to consume less, to drive less, to eat less meat, to consume less water, etc.). This logic of loss is one of the biggest problems in the discussion about sustainability today: "Turning this logic around, changing the perspective, is therefore the most important implementation of the UN Sustainable Development

Goals. And this is first and foremost a cultural challenge" (Zimmermann 2018: 17).

The UN 2030 Agenda from 2015 (see link 30) defines sustainability policy as a cross-cutting task of everyone, including governments, civil society and every individual. The development agenda does not formulate culture as a goal in its own right, but "the importance of culture as an integral part of implementing all 17 Sustainable Development Goals is undisputed" (Bärwolff & Schneider 2019: 19).

Because humans are pleasure-seeking creatures, the permanent renunciation attitude will not work with a majority.

> The topic of sustainability must be dealt with culturally. If we succeed in this, it will no longer be renunciation that comes first, but profit. The economic gain, because sustainable management has long been a market and economic factor. The ecological gain, because the preservation of our natural basis of life is essential for our survival. The social gain, because a sustainable society is oriented towards the common good. The social gain, because living in a world in which nature and culture are permanently compatible is the prerequisite for a good life (Zimmermann 2018: 17).

It is generally agreed that the discourse on sustainable development must also include the field of cultural policy (see, among others, Schwencke 2003; German Bundestag 2008; Rydzy & Griefahn 2014): "A debate on sustainability with reference to the arts and the integration of cultural policy actions remains a major challenge for the 21st century." (Schneider 2014: 22). Only a rethinking in all areas of life can initiate successful sustainable development, and cultural approaches can promote this change of perspective:

> Current debates show that people are only able to tackle problems of sustainability when there is a change of cultural attitudes, because sustainability demands a massive rethinking of all areas of our lives, our established ways of thinking and behaving. Indeed it goes to the root of all our basic values (Schneider 2014: 20).

A transformative cultural policy strives for stable sustainability and resilience. This corresponds to the EU criterion of a minimum ten-year development strategy for ECoC candidate cities, which have to deal with continuing and accelerating change (in the above-mentioned categories). However, the most comprehensive evaluation study of the ECoC initiative to date, the so-called Palmer Report, still attested in 2004 that the effects on the Capitals of

Culture to date did not extend much further than the programme year (see Schwencke 2015: 115; Palmer 2004). In the meantime, however, not least due to the revised and expanded legal basis of the initiative, those responsible for the programme have become more aware that the sustainable effects that can arise through the self-chosen transformation process within the framework of an ECoC (bid) are the actual essence of the programme. The ECoC year is thus now seen only as part of the long-term development strategy, which is conceived far beyond the programme year. In the case of the applicant cities, this also concerns a tailor-made "Plan B", i.e. an alternative scenario if their bid is not successful.

In order to ensure the sustainable planning of a transformation process, stable governance structures are needed (see above). Thus, both the theory of cultural governance and the approach of transformative cultural policy form the theoretical framework of this work. Wolfgang Schneider formulates the interlocking of both theories as follows: "The key concept of sustainability can only prove successful if it is built on a global, interdependent foundation, based on partnership. It must therefore also include cultural policies, which in turns requires a new governance: a good governance for cultural policy" (Schneider 2014: 22).

4. The intrinsic cultural logic of cities

4.1 The theory of the intrinsic logic of cities

For a long time, urban sociology was dominated by the theoretical tradition from the beginning of the 20th century, which was clearly influenced by Georg Simmel and Max Weber.[15] According to this, a "city" – in contrast to "rural areas" – was a laboratory for social processes, and a place "where society appears in its structure and its conflicts" (Siebel 1987: 11). Later, from the 1980s onwards, the focus was on small-scale social processes, for example, in the district, neighbourhood or milieu. The scientific interest here was in the spatialisation of social specifics, for example, the lifestyles and particularities of certain city districts.

The theory of the intrinsic logic of cities of Martina Löw (2008), on the other hand, is an attempt to understand the city with its social phenomena as a whole. While cities are becoming increasingly similar in the course of globalisation, the theory of the intrinsic logic of cities focuses precisely on the local cultural identities and particularities of the locations that distinguish them. It is a truism that cities and their inhabitants differ – but, until now, science had not yet explored the question of why this is so and what conclusions can be drawn from it.

"The continuous substitution of the research topic 'city' by 'society'" (Löw & Berking 2008: 8) in the field of urban research is overcome by the theory of the intrinsic logic of cities, in that the city itself becomes a social fact. The intrinsic logic approach also contradicts the prejudicial view that cities are too diverse and too complex to be considered and compared scientifically as a whole. On the contrary, the approach focuses on a locality's own identity beyond the development of urban structures homogenised by globalisation and further postulates that the structure of a city has an effect on its inhabitants.

---

15 Among others, by Georg Simmel's essay "Die Großstädte und das Geistesleben" from 1903 (Simmel 2006) and Max Weber's treatise "Die Stadt" from 1921 (Weber 2009).

A comparative study by the research team comprising Ian Taylor, Karen Evans and Penny Fraser (published in 1996) on the two British cities of Manchester and Sheffield can be seen as the starting point of the theory. In their analysis of the different development paths of the two cities in terms of everyday practices, the following becomes clear: "It is still sensitive, even in these globalising times, to recognise local cultural differences between cities [...] and to treat them as having a sociological significance and continuing cultural provenance and impact" (Taylor et al. 1996: xii). The study compares the local practices of the two northern English cities, which both have to deal with structural change in the post-industrial age. The study, which analyses a broad spectrum of aspects, from the organisation of local public transport to shopping facilities and residents' feelings of belonging, substantiates the suspicion that "cities function much more fundamentally in an idiosyncratic way" (see Löw 2008: 39). According to Löw, this concerns the basic structures of a city that affect all areas of life, and which also do not have to be unique, but certainly can be found in similar structural patterns in other cities. The intrinsic logic of a city, according to Löw, cannot be traced back to individual actions like an image campaign:

> Neither the mayor nor the advertising expert nor the directors of the banks succeed in determining the experiential space of a city on their own. In routinised and institutionalised practice, condensation and demarcation, constructions of uniqueness and unity can be understood as a site-specific and thus differentiated production of meaning in which very different social groups participate. In their actions, each group is to be understood both as a co-producer of intrinsic urban logic and as a product of city-specific meaning (Löw 2008: 45).

Using a scientific comparative methodology, differences and similarities between cities can be systematically explored. After all, poverty, wealth and cultural life are phenomena with locally specific differentiations, meanings and perceptions. Winston Churchill's 1944 quote serves as an example of anecdotal evidence: "We shape our buildings and afterwards our buildings shape us" (see link 20). Churchill's thesis, which he related to the rebuilding of the House of Commons, which had been destroyed by the German Luftwaffe's attacks on Britain during the Second World War ("The Blitz"), shows the extent to which the intrinsic logical structure of a place also affects its inhabitants.

This is because the intrinsic logic, i.e. the "concealed structures of cities as locally practised, mostly tacitly effective pre-reflexive processes of meaning formation together with their physical-material inscription" (see link 21), influences the actions of individuals and groups, i.e. in Bourdieu's sense, their habitus. The intrinsic logic thus describes "a complex, inter-woven ensemble of knowledge stocks and forms of expression based on rule-governed, routinised forms of action stabilised by symbolic as well as material resources" (Löw 2009: 55). Examples of this are city-specific practices such as linguistic turns of phrase, social norms (such as special forms of politeness) or generally accepted values that are reflected, for example, in the way in which the history of the location is addressed.

The intrinsic logic approach analyses the basic socio-cultural specificity of cities and can lead to insights and tools for identifying and promoting local specificities. However, because the intrinsic logic of a location depends on very many (historical, geographical, financial, etc.) factors and is formed by a continual process of cultural external- and self-definition of urban iden-tity, an interdisciplinary and comparative research approach is required.

4.2 The intrinsic cultural logic of ECoC locations

The "intrinsic cultural logic of cities" is based on Martina Löw's theory and is adapted for the cultural sector. It is based – as is this entire work, see above – on a broad understanding of culture according to the UNESCO Declaration on Cultural Policies (Mexico City 1982). In addition to the various art forms such as literature, music, theatre and visual arts, it also encompasses ways of life, fundamental human rights, value systems, traditions and beliefs and is based, among other things, on the principle that cultural identity and cultural diversity are inextricably linked to each other.

In times of global competition between locations and the re-location of cities through post-industrial structural change, the "idea of a local-specific, idiosyncratic reality of cities" (Berking & Löw 2008: 7) is more significant than ever. In this context, the intrinsic logic approach has opened up a spec-trum of new criteria for comparing cities, which can serve the self-assurance of the inhabitants and the external presentation equally.

The ECoC initiative can act as an occasion and a means to promote and contextualise urban development projects, as well as to reflect on the individual specificities of the location and their significance for the future of the city (see Jacobsen 2018: 11). This catalytic effect that candidate cities rely on, even if they do not win the title, leads to the overarching EU criterion of being a model in terms of fostering "the contribution of culture to the long-term development of cities in accordance with their respective strategies and priorities" (see link 5).

The approach of the intrinsic cultural logic of cities can help to analyse the cultural specifics of a location in the context of a bid to become a Capital of Culture. What cultural offerings are there in the city? What is the reception behaviour of the various city residents? What cultural policy decisions are made in local politics, and what effect do they have? What does the cultural institutional structure look like, and what networks or governance structures exist in this area and beyond? To what extent does the cultural sector of a city contribute to its overall logic? And looking to the future, for example: How can further cultural anchor points be realised in the ECoC process with the consent of the population?

If an ECoC (or a candidate city) claims to give the city a new image, as was the case with RUHR.2010 and MP 2013, the following question arises according to the intrinsic logic approach: "What adjustments have to be made?" (Martina Löw, in a background interview with the author). This requires an unbiased examination of the city's own structures of its intrinsic logic. This can be helpful in the case of candidate cities (for example, locations that are similar according to formal criteria such as size, location or development history) vis-à-vis the other cities competing.

In reality, however, such analyses are usually lacking at the beginning of a bidding process as a Capital of Culture, although they can be well-founded and point the way for the further path of the bid. In the initial phase, when the investigation would make the most sense, there is usually very little cooperation with universities, research institutions or other bodies that could take on this task.

Over the years, certain codes have emerged among the ECoCs and candidate cities in their approach to participation in the initiative; there are a few "grey eminences" who act as spin doctors, and alongside them a relatively small circle of advisors who move from ECoC to ECoC. The already frequently voiced accusation of a "travelling circus" refers to the inevitably comparatively similar approach of these actors at the various

locations – even if an opposite approach, namely one tailored to the location, is frequently and forcefully proposed. But it can be problematic to
adopt elements from the Capitals of Culture that are considered successful,
such as RUHR.2010: citizens will notice sooner or later whether an ECoC
strategy "fits" or not, and support, reject or ignore it accordingly. It is precisely here that the intrinsic cultural logic approach can prevent, through
scientific analysis, the individual strengths, challenges and potentials of a
city from being ignored by a schematised approach.

The visual language used emblematically by the two ECoCs studied
in this paper – the MuCEM for Marseille, the Zeche Zollverein for
RUHR.2010 – alone reveals their intrinsic-logical self-presentation. In this
sense, the intrinsic logic approach "starts with the respective branding and
image strategies of the cities on the assumption that conclusions about
their intrinsic logic can be drawn from the way a city tries to present its
own particularities" (Löw 2008: 187). However, these examples show the
limitations of the intrinsic logic theory: Its validity should rather be understood as a tendency, since the complex object of investigation, the "city"
as a whole, always remains ambiguous. The MuCEM, for example, is not
a municipal museum, but a national museum conceived from Paris and,
as will be shown, was intended rather to promote national interests (see
chapter 8). The MuCEM thus shows that the "typical ways for cities to
stage their own" (Löw 2008: 241) must be considered in a differentiated
way, especially with regard to the actors involved.

5. The research instrument: Expert interviews

5.1 State of research and methodological approach/Grounded Theory

Although the relevant literature on the ECoC initiative has become increasingly extensive over the years, it is so diverse that no systematic recurrence can be discerned in the approaches taken by the publications. The literature is heterogeneous and partly interdisciplinary (e.g. geography/urban development, political, cultural and European studies, tourism research). Due to the different approaches in the literature, only a limited comparison is possible. This is also the result of the rotating nature of the programme and its relatively short duration at one location. Compiling a systematic overview of previous publications on the subject of the ECoC initiative would be a further area of research.[16]

There is little literature from a European, i.e. higher-level perspective on the development of the ECoC initiative. Obviously, this is related to the fact that it is still primarily understood as a municipal, regional or even national event, and not as a European one (see Jacobsen 2009: 47). The increasing number of Bachelor's and Master's theses on the topic also deal predominantly with the individual locations of the initiative and rather less with overarching development processes or systematic issues.

The ECoCs MP2013 and RUHR.2010 are located in the two largest EU member states; presumably for this reason alone, a relatively large number of studies have been published on them compared to other ECoCs. However, since the two programme years took place quite recently, there are only a few publications on their sustainable development.

Field access to primary and secondary materials on the ECoC topic can be described as limited. Even academics and other experts who have been working on the ECoC initiative for a long time are dependent on

---

16 See inter alia Baier & Scheytt (2010: 158): "In addition, it is desirable that the effects, modes of operation, and history of the European Capital of Culture are scientifically processed even more intensively and continuously."

information and assessments, but also documents, to which the only access is individual expert discussions.

On the one hand, this insufficient field access is a result of the fact that there is still no archive of the ECoC initiative. The European Commission is holding back on this, as it generally plays a relatively passive role in the entire ECoC series (see Jacobsen 2017: 7). For example, the documents available on the ECoC topic on the homepage of the COM are only overview-like, but not very extensive (see link 13). The only archive that exists on the topic so far is located in the Archive for Social Movements in the House of Ruhr History (Archiv für soziale Bewegungen im Haus der Geschichte des Ruhrgebiets) in Bochum. It contains extensive documentation material on RUHR.2010; however, this material was only partially processed and accessible when the research for this paper was conducted. Numerous files were blocked until 2020 or are still blocked or not yet accessible for capacity reasons. However, the municipal archives of the Ruhr region have been cooperating more closely since the ECoC year. Jürgen Mittag sees this as a positive example of inter-communal networking as a result of RUHR.2010. Even if there has not been a complete reorganisation of the archives, the ECoC year has led to an improvement in the already existing contacts between the Ruhr archives. This applies not only to the municipal archives but also to the supra-municipal or specialised archives, such as the Archive for Westphalian Economic History, the industrial archives of Krupp and Thyssen, etc., which are now assigned to the Ruhr Regional Association (Regionalverband Ruhr, RVR) (Jürgen Mittag 2016: in interview).

There are no archives on MP 2013 or the other ECoCs.

Individual private consultants or companies advise the municipalities during the bidding process or later during the implementation of the ECoC year. A Europe-wide institution that is equally available to all ECoCs and candidate cities in an advisory and public capacity, even if only by providing the documentation, has not yet been established. As a result, any city considering entering the bidding process has to muster considerable resources to compensate for this deficit if it wants to explore its chances and risks as a potential ECoC.

For these reasons, the own data collection through qualitative interviews with experts represents a necessary continuation of the survey methods for ECoC research (see Flick 2006: 97 f). In contrast to the quantitative approach, the qualitative survey method inductively develops correlation theses in the course of the investigation, i.e. theses that assert that various characteristics that have been worked out show interdependencies. The qualitative-explorative research design thus attempts to arrive at extended analytical results about processes, structures and their effects at the two locations.

The approach of this work is based on the Grounded Theory, which was developed in the mid-1960s by Anselm Strauss and Barney Glaser in the tradition of pragmatism and symbolic interactionism. The term "Grounded Theory" already indicates that empirical facts form the starting point of research and theory generation. The "Grounded theory can be described as the classical, theory-discovering qualitative method", (Brüsemeister 2000: 189) which is used to develop a theory that is inductively derived from the generated material collected to answer a question (see Strauss & Corbin 2010: 7–9). The theory is derived from an ongoing comparative analysis. It does not describe reality as in pragmatism or positivism, but represents a perspective – and fallible – interpretation of the collected data. However, the understanding of the Grounded Theory underlying the present work must be distinguished from approaches of constructivism, in which key social processes assume a central position for theory construction. According to the approaches of the Grounded Theory, the results are not incontrovertible theories, but are subject to changes and modifications, just like the examined reality (see Lueger 2007: 195 f).

According to the Grounded Theory, the empirical procedure should be divided into three steps:

1. *Theoretical Sampling* or *Selective Sampling*, in which the analysis is driven by the ongoing data collection oriented towards the results. The ongoing interpretations of the surveys thus provide the criteria for the progress of the investigation, which is intended to corroborate the emerging theory.
2. The *Coding Paradigm* describes the "translation" of data and the naming of overarching concepts, as well as their explanation and discussion. Indicators related to the research question are sought in the data, from which preliminary concepts are derived that always relate directly to the empirical material to be interpreted. During the evaluations, the concepts become more and more differentiated, numerous and

abstract, and are finally grouped into superordinate categories. Either independent categories (*open coding*) or interlinkable categories (*axial coding*) are identified. This involves asking about conditions or interactions between actors/strategies and tactics/consequences, with *axial coding* identifying temporal and spatial relationships, cause-effect relationships, means-purpose relationships, and argumentative and motivational relationships. Another possibility is *selective coding*, which builds on already existing categories (see Böhm 2000: 475–485).

3. *Writing Memos* means that in the process of coding, connections are made between the categories, which finally lead to the formation of hypotheses. The hypothetical results must be checked again and again using new data material in a deductive procedure (see Glaser 1978; Glaser & Strauss 2009; Strauss & Corbin 2013).

The process of analysis is thus triadic and circular at the same time in the sense of the hermeneutic circle. Conclusions are reached through a combination of inductive and deductive procedures. New data that are suitable to verify the previous conclusions are always coded. "It is always empiricism against which a theory has to prove itself and to which theory always returns as the last instance" (Hildenbrand 2000: 36). The Grounded Theory should "in some way identify action and change or the causes of little or no change" (Strauss & Corbin 2010: 100). In this respect, the study of transformation processes is a central approach of the Grounded Theory. It offers tools to identify the "fundamental processes that bring about change" (Hildenbrand 2000: 32). These processes "are influenced by change and in turn influence change, i.e. they bring it about" (Hildenbrand 2000: 32).

However, a criticism is that the method demands that the researcher approaches his object of investigation as unbiased (and unread) as possible in order to be more open to theory generation. At this point, however, the Grounded Theory disregards the fact that even the initial selection of data is based on the researcher's individual research interest and prior understanding:

> Findings about social phenomena do not "emerge" of their own accord; they are constructions of the researcher from the beginning. The idealisation of the researcher's "impartiality" and the idea of a "direct" grasp of social reality, which can occasionally be found in qualitative methodology, are thus epistemologically untenable (Meinefeld 2000: 269; see also Meinefeld 1995: 287–294).

In fact, selection decisions are made in the above-mentioned three stages of the entire research process (i.e. during data collection/selection of material and during the interpretation of data, as well as during the presentation of research results).

5.2 The necessity and importance of expert interviews

The far-reaching importance of expert interviews for research practice is evident: in the social sciences there is hardly any field in which they are not part of the core of research practice. Especially for ECoC research, interviews with experts are indispensable, since the available literature is limited (see above). Here, the expert interview is the research instrument of the qualitative research method, which aims to establish causal connections. By concentrating on a few cases (small-n study or in-depth study), the aim is to achieve a gain in "holism and complexity" (Kuckartz et al. 2009: 13).

In this study, the expert interview is used to gain information in relation to the research questions of the study. At the same time, subjective interpretations are recorded and placed in context. The qualitative interviews were always guided, i.e. most of the questions were prepared in advance, although the interview process was flexible in the formulation of questions, questioning strategies and the sequence of questions.

The appendix of the dissertation, i.e. the fully transcribed expert interviews, is not included in this publication.

The experts were selected on the basis of their status as professional function holders or process observers. Because being an expert is never a personal ability or characteristic, but always a subjective attribution, criteria for the selection of the experts were defined in advance: they had to have a recognised reputation in connection with the ECoC initiative and had to have been involved with the topic for years. In addition, as programme managers, participants or outsiders, they had to provide qualified assessments from both internal and external perspectives. The requirement to select interview partners who are as knowledgeable as possible thus presupposes the author's prior knowledge of the case under investigation, which at this point represents a sensible modification of the Grounded Theory and raises the standard of the research results.

The experts were not selected according to the *most similar case design* nor according to the *most different case design*, which is often used for comparable research objects, but according to the *diverse cases method / multimethod research* (according to Seawright & Gerring 2008: 296 f). This approach has the advantage of openness to both convergent and divergent tendencies in terms of content and the goal of reflecting a broad spectrum of perceptions.

In the sense of the triadic circle of Grounded Theory, different inter-view partners and documents were selected in the initial phase in order to obtain data that reflect as broad a spectrum of the research questions as possible. Later, on the basis of coding and formulating memos, both the selection of interview partners and the guidelines were expanded and changed in order to in turn confirm or differentiate the categories and theories already (provisionally) developed. This is the essential difference from other techniques of sampling: In theoretical or selective sampling, the critical examination of the case or the selection of experts is already part of its construction. The method also has the advantage of guaranteeing the central scientific research criteria through an adapted and broadened data basis:

- objectivity / confirmability of qualitative studies
- reliability / auditability
- internal validity / credibility / authenticity
- external validity / transferability / fit
- usefulness / application / action orientation

(see Steinke 2000: 320 and Kelle 1997: 277 ff)

With its claim to developing theories of medium scope, the Grounded Theory justifies the scope of the empirical material, i.e. the point of the-oretical saturation, as soon as the information from the various experts consulted is repeated and no longer represents a significant extension of the data. However, because the method is fundamentally circular in structure, theoretical saturation can only be justified to a limited extent according to the Grounded Theory and must be adapted not least to the feasibility of the research project. Also, the researcher would move further and further away from the initial questions if the circular procedure were not terminated by the defined theoretical saturation.

5.3 The interview partners

The interviews conducted for this paper aimed to supplement the primary and secondary literature on the ECoC topic with expert assessments of the two sites RUHR.2010 and MP 2013. This involved a broad spectrum of perspectives, which is why the experts came from very different areas between programme responsibility and process observation. In the spirit of the Grounded Theory, the interviews evolved during the implementation phase as the author's level of knowledge and reflection improved.

The guided interviews were conducted with the following experts:

RUHR.2010
From the fields of politics, media, associations, science and with programme managers:

– Gerald Baars
  Studio Manager at Westdeutscher Rundfunk (WDR)
– Jens Dirksen
  Head of Feuilleton at the Westdeutsche Allgemeine Zeitung (WAZ)
– Jürgen Fischer
  Head of the Department for Culture and Sport at the Regionalverband Ruhr (RVR); former Manager for International Relations at RUHR.2010
– Prof. Dr. Jürgen Mittag
  German Sport University, Cologne
– Prof. Dr. Oliver Scheytt
  ECoC Consultant, former Managing Director of RUHR.2010 GmbH
– Karl Schultheis
  Member of the Landtag of North Rhine-Westphalia since 2005, Chairman of the Committee for Culture and Media (2012–2017)

MP 2013
From the fields of art and culture, science and with programme managers:

– Alain Arnaudet
  Director of La Friche la Belle de Mai
– Antonia Blau
  Staff member of the Goethe Institut in 2013

- Jean-François Chougnet
  Director of the Musée des civilisations de l'Europe et de la Méditerranée
  (MuCEM)
- Prof. Dr. Ulrich Fuchs
  Deputy Director of MP 2013
- Pia Leydolt-Fuchs
  Consultant for ECoCs, based in Marseille
- Gilles Suzanne
  Aix-Marseille University (AMU)

In addition to the interviews, numerous background discussions were conducted, for example, in the context of participant observation at cultural events or conferences at the two locations or when visiting the cultural institutions there. The conversations with cultural actors and consumers were mostly situational and tended to be brief, so that they conveyed an impression, often in the sense of anecdotal evidence.

6. RUHR.2010

6.1 Transformation in concrete terms: Dispositions, conceptions and intrinsic cultural logic

With its bid for the ECoC title, RUHR.2010 initially formed part of a series of past Capitals of Culture that also wanted to use the ECoC title as an occasion and framework for an urban development process at a former industrial site (Glasgow 1990; Rotterdam 2001; Genoa 2004; Liverpool 2008). The cultural policy disposition for the ECoC RUHR.2010 was a polycentric conglomerate with a lot of "parochial thinking" (*"Kirchturmdenken"*). This means that despite the spatial proximity, each municipality in the Ruhr has its own cultural policy administrative structure without being linked to the cultural administrations of the surrounding cities. The aim of the ECoC RUHR.2010 was to advance urban and regional development through a cooperative and innovative cultural policy in the "Metropole Ruhr". After all, the political fragmentation in the various administrative units of the Ruhr region (four counties [Landkreise], 11 independent cities, three administrative districts [Regierungsbezirke] and two regional associations [Landschaftsverbände]) presented the participating 53 cities with considerable organisational challenges. At first, even Oliver Scheytt, the managing director of RUHR.2010, doubted that the Metropole Ruhr really existed: "We simply claim that ourselves. This is a kind of surprise attack" (quoted in Graalmann 2010). But the Ruhr region is characterised precisely by its polycentricity, to the extent that it is hardly conceivable that there will ever be a genuine Metropole Ruhr in which the city and, above all, administrative boundaries are fluid (see Kießlinger & Baumann 2016: 115).

The programmatic motto of the ECoC RUHR.2010 was "Change through Culture – Culture through Change" and was intended to trigger the desire for culturally influenced urban development in the participating municipalities. This cultural policy objective – broadly conceived, both in terms of its regional scope and its understanding of culture – encompassed, on the one hand, the physical substance by transforming former industrial sites into cultural venues. But the motto equally referred to a readjustment of the self-perception of the inhabitants of the Ruhr region (internal perception) as well as to an image change or a new city branding of the region (external presentation). The motivation of a transformative cultural policy convinced the national selection jury, which recommended to the European Commission the location "Essen for the Ruhr" from a total of 16 German candidate cities for the ECoC title. "Not least, the transformation process of the Ruhr region towards a metropolitan region, which continues to this day, was the driving force behind the Ruhr region's successful bid to become RUHR.2010 – European Capital of Culture" (Kießlinger & Baumann 2016: 115). The designation as ECoC took place in 2006 and, even before that, in the bidding process, the course was set for a long-term transformation strategy that would last far beyond the ECoC year.

RUHR.2010 did not have to start from scratch like other Capitals of Culture, but was able to build on the network structures of the International Building Exhibition Emscher Park (IBA Emscher Park), which ended in 1999 after a ten-year run. The IBA Emscher Park had previously been decided upon by the North Rhine-Westphalian state government "and de facto imposed on the Ruhr region" (Betz 2011: 329). The last years of the IBA Emscher Park saw the first regional cultural projects, which were followed by the Ruhrtriennale from 2002. But in contrast to the IBA, RUHR.2010 focused less on material renewal and more on changing the image of the Ruhr region both internally and externally. This change was in keeping with the spirit of the times, because apparently there had also been a desire among the population for some time to change the image of the "Ruhri" with a coal-smeared face.

Since the beginning of the 21st century, the number of cultural and music festivals in the Ruhr region with supra-regional appeal has grown, e.g. the *Extraschicht* since 2001 and the *Loveparade* with hundreds of thousands of visitors since 2007. Even before the ECoC year, in 2007, the "Initiativkreis Ruhrgebiet", an association of the 60 largest Ruhr corporations and co-partners of RUHR.2010 GmbH, demanded in a strategy paper that in order to "raise the international profile of the Metropole Ruhr [...] a major event" such as a world exhibition or Summer Olympic Games should be held at regular intervals (Initiativkreis Ruhrgebiet 2007: 27 f). This already referred ex ante to RUHR.2010 as a trend-setting example for future major events in the Ruhr region (Betz 2011: 333).

RUHR.2010 was undoubtedly a major event. It was by far the largest ECoC at that time, and it adorned itself with superlatives regarding its geographical extent, its budget and its level of awareness. RUHR.2010 was therefore very self-confidently labelled a "millennium event" or "German New York" by those responsible for its programming (Fritz Pleitgen in the Süddeutsche Zeitung, see link 23). According to the organisers, the programme had 10.5 million visitors and thus achieved a considerable increase in tourism in the region. The greatest success of the ECoC year can be seen in the improvement of the image of the Ruhr region, so that the motto "Change through Culture" actually had a demonstrable impact: according to a Forsa survey commissioned by the Initiativkreis Ruhr,[17] in 2008 only 2% of respondents thought of culture and the cultural offerings when they heard the term "Ruhr region"; in 2012 the figure was 12%. In 2012 only 34% thought of coal; in 2008 it was 53%. In addition, 3% of respondents mentioned "football" in 2008; in 2012 it was 7%. "Culture thus beats football in this nationwide survey in the chain of associations with the Ruhr region," the organisers sum up (RUHR.2010 2003). The concept of adapting the Ruhr's own intrinsic cultural logic to the region's structural change with a new narrative has borne fruit through RUHR.2010, at least according to the Forsa survey.

6.2 Perspectives of cultural governance in the aftermath of the ECoC RUHR.2010

The ECoC RUHR.2010 is a good example of how the output of policies in the governance sense is a complex structure of cooperation and negotiated solutions and that there is no single top-down political control. For RUHR.2010 GmbH did not centrally organise the more than 300 projects with more than 2,500 events in the participating 53 cities. Only a few events, such as the lighthouse project "Still-Leben", were organised by RUHR.2010 itself, whereas the majority of events were cooperation projects or external events for which cultural institutions or individual artists had applied. There

---

17 508 residents of the Metropole Ruhr and 1,002 people outside the Ruhr area in the Federal Republic of Germany were surveyed in 2008 and 2012: see RUHR.2010 (2012).

were considerable differences in the extent of cooperation, also in terms of subsidies. The "Local Heroes" project is the clearest example of the decentralised governance structures in RUHR.2010: in each week of the ECoC year, another of the 53 cities presented itself and organised its own cultural events within the overall programme. Essen, as the banner bearer, remained present throughout the year (see Mittag 2019: 200).

Subsequently, some elements of the cultural governance established on the occasion of the Capital of Culture were continued or even expanded. These include the still successful association of 20 art museums (*RuhrKunstMuseen*) and the network of stages in the Ruhr region, which finally led to the *RuhrBühnen* (11 stages in nine cities) in 2015. The joint use of various cultural actors within the new design concept of Zeche Zollverein should also be mentioned here, as well as the inter-communal sing-along festival*!Sing – Day of Song*, which continues to this day. When the latter event took place for the second time (2012), only two-thirds as many people took part. Nevertheless, the RVR was committed to continuing the big sing-along concert because it was considered "identity-forming for the local population" (Karl Schultheis 2016: in interview).

After the ECoC year, 4.8 million euros were made available annually and for an unlimited period as sustainability funds (half from the RVR and half from the federal state of North Rhine-Westphalia), which are topped up in part by the cities' own funds and regularly also by tourism funding from the federal state of NRW. The funds are mainly allocated to *Kultur Ruhr GmbH*, which finances events and activities under the title *Urban Arts Ruhr (Urbane Künste Ruhr)*, and to *Ruhr Tourismus GmbH*, which, among other things, continues the above-mentioned *Day of Song*. Sustainability funds are also used for the annual *Ruhr Area Cultural Conference* and the *European Centre for Creative Economy (ecce)*. *Emscher Art (Emscherkunst)*, initially conceived as a triennial, will now continue as a permanent *Emscher Art Trail (Emscherkunstpfad)*. The initiative *An Instrument for Every Child (Jedem Kind sein Instrument)*, which is intended to enable primary school children to take part in music lessons across the board and was founded for the Capital of Culture with state and federal funds, is continued in a modified form by a foundation.

The continuing initiatives and projects are each based on their own governance structures. The organisational framework is provided by the

RVR with its culture department and the Conference of Cultural Directors (Konferenz der Kulturdezernenten). In addition, there is the annual *Ruhr Culture Conference (Kulturkonferenz Ruhr)*, where the cultural actors of the Ruhr region can exchange ideas and network.

In order to ensure the sustainability of what has been achieved, no central coordination office has been established; instead, in the sense of cultural governance, it is now the responsibility of various existing organisations that are financed to maintain and expand the networking that has been established in the cultural sector. This, however, carries the risk of becoming "too fixated on the institutions that are there" (Jens Dirksen 2016: in interview), which may be less open to innovative cultural and cultural policy approaches than new forms of organisation.

Nevertheless, Karl Schultheis' assessment that the sustainability architecture of RUHR.2010 is unique among the previous Capitals of Culture is justified: Never before has an ECoC secured the maintenance and continuation of what has been achieved in such a way, both structurally at the governance level and financially through corresponding contracts with the political levels involved.

However, even with the improved governance structures, not all of RUHR.2010's goals could be achieved. For example, the districts around the Zeche Zollverein are still characterised by unemployment and low incomes. The hope that the new utilisation concept for the huge area of the former colliery would create numerous jobs has not (yet) materialised. Many of the residents are also critical of the Zollverein World Heritage Site and feel little attraction to the place of supposed high culture (see Kießlinger & Baumann 2016: 117 f). However, the development of the colliery site has not yet been completed, and the users of the site still intend to improve the identification and participation of the population.

Although many cultural practitioners wished for more follow-up projects after the 2010 programme year, the sustainability of RUHR.2010 and its indirect influences through the successful establishment of network and governance structures can be seen as positive. This certainly also had an impact on the "European Green Capital" (see links 34 and 35), which took place in Essen in 2017 and brought the city another tourism high after RUHR.2010.

6.3 The ECoC as a location factor for city marketing

For the ECoC, institutions such as the *Zeche Zollverein* or the *Dortmunder U – Centre for Art and Creativity* established themselves as creative industry centres with supra-regional visibility. Round tables were set up, bringing together art and culture professionals with entrepreneurs and local politicians. An example of regional cultural governance is *ecce*, which acts as a network of public and non-public partners. It has set itself the task of financially supporting artists and creative professionals within the funding programme *Kreativ.Quartiere Ruhr* when they realise project ideas together with other local partners that contribute to culture-based urban and neighbourhood development. However, Birgit Mandel is critical about the fact that these public funding programmes are still little known and that there have been hardly any evaluations of the programmes' effects so far (see Mandel 2017: 20). Moreover, the main shareholder of *ecce* is the *Business Metropole Ruhr (BMR)*, whose main task, according to its self-description, is to market the Metropole Ruhr as a business location.[18] There is therefore no question of promoting a cultural landscape with an understanding of culture as a (purpose-) free space for social forms of expression.

The ECoC year 2010 undoubtedly represented a peak for cultural actors in the Ruhr region. Whether the cultural funding landscape as a whole has improved as a result of RUHR.2010 was assessed differently in the expert interviews conducted for this work. Karl Schultheis answered this question in the affirmative and justified this with the increase in EU cultural funding, which could only be applied for jointly by cooperation partners from several countries. From his perspective, much of the cooperation only came into being because of RUHR.2010 (Karl Schultheis 2016: in interview). Jens Dirksen contradicts this and claims that the cultural scene was not the focus from the beginning, but that it was only instrumentalised as a vehicle

---

18 "Business Metropole Ruhr GmbH (BMR) develops and markets the Metropole Ruhr as a business location. It bundles the economic interests of the 53 cities in the region. The aim of BMR's work is to increase the competitiveness of the Ruhr region. This includes implementing and continuously developing an overall economic strategy for the Metropole Ruhr and promoting the image of the business location as a high-performance and innovative region both nationally and internationally." (See self-description on the BMR homepage under link 12.)

for marketing the Ruhr region. As a result, in his opinion, RUHR.2010 has not changed anything in terms of cultural policy (Jens Dirksen 2016: in interview). He is correct in that since the ECoC year there has been a close link between culture and tourism promotion (see below), which suggests a profit-oriented relation of the "success" of the cultural landscape – reduced to an economically oriented profitability – to the capital invested.

"From myth to brand" is how RUHR.2010 managing director Fritz Pleitgen summed up the ECoC's task of dealing with the structural change of the Ruhr region (Pleitgen 2010: 6). This formulation definitely resonates with a utilitarian understanding of culture as an instrument to achieve economic goals. Jens Dirksen and Jürgen Mittag attest to this instrumentalisation of RUHR.2010 in an interview. It is also expressed in the fact that both the museum and the theatre network are located at Ruhr Tourismus GmbH, whose main interest is the promotion of tourism per se, above the facilitation of cultural open spaces. The connections shown are usually not well-known to the public and clearly contradict the goals that are usually promoted in cultural policy discourse.[19]

6.4 Preliminary conclusion I: Criticism and lessons learnt

Some points of criticism that apply to many Capitals of Culture are also applicable to the conception of RUHR.2010. These include the perceived insufficient involvement of local artists in the programme. Apparently, deciding where and to what degree local artists play a role in the ECOC programme is a difficult challenge for all ECoCs. On the one hand, there is a need for renowned artists who cultural tourists already know from other

---

19 E.g. "We need to promote art and culture without purpose – only then do they generate meaning." (Brosda, Carsten in Burghardt: 2018) or "For a holistic, integrative urban development with a community-first perspective that does not primarily follow an economically narrowed perspective of the 'city as a corporation', this means: The consciousness of city can be developed positively through culture." (Könneke 2011: 120) or "Is it about the functionalisation of artists again? Are cultural actors only a means to an end? Is this the cultural policy answer to globalisation? After all, cultural exchange should be experimentation, bringing artists together a laboratory, possibly to promote creativity, possibly also with the risk of failure." (Schneider 2008: 27); see also chapter 1.2.

major exhibitions at famous sites and who are strategically deployed to act as a "tourist magnet". For the cultural tourist, like the package tourist, also likes to travel to places where he or she finds familiar things again (see Jacobsen 2018: 15). On the other hand, however, EU regulations demand that the programme be novel-innovative, locally anchored, participatory and authentic. This conflict manifested itself most recently in the large-scale project *Elf Fonteinen* of the ECoC Leeuwarden-Friesland 2018 (LF 2018). Here, 11 new artistic fountains were created in 11 cities in Friesland, in memory of the tradition of speed skating that used to take place in winter between the 11 cities on the frozen rivers and canals. The 11 fountains were designed by renowned international artists such as Stephan Balkenhol and Mark Dion. This angered the local art scene: "Why spend so much money on the usual elitist travelling circus of contemporary art, when there was also enough creative potential in the region?" (ibid.). The ECoC organisation LF 2018 officially responded on its homepage: "For *11 Fountains* we chose international artists because we simply wanted to have special, beautiful fountain artworks. [...] By choosing international artists, we dare to dream even bigger and stand out from the crowd. That is un-Frisian, if you will. But with that, it is also a sign of this new phase" (see link 24). It is doubtful, however, whether this statement helped to appease the local artists and cultural actors (see chapter 8).

The example makes it clear that normative aspects are always immanent in the decisions taken within the ECoC framework, which are first and foremost taken up by the population and can be understood *pars pro toto* as the supposed character of the entire ECoC strategy (see chapter 2.4). Also in RUHR.2010, local independent actors did not feel sufficiently involved in the ECoC programme, which led to "frustration effects" (Jürgen Mittag 2016: in interview). Jürgen Fischer, who was responsible for the programme, blames the artists themselves for this:

> That also had to do with the fact that the Independent Scene, unlike others, only understood relatively late how the Capital of Culture works. That we did not provide individual funding, but always tried to initiate networks and cooperation. That also worked with the museums and theatres. In this respect, I would not really accept this criticism (Jürgen Fischer 2016: in interview).

Another way of looking at the conflict, which is certainly not wrong, would be that the ECoC organisers apparently did not communicate their participation strategy sufficiently to the independent sector. Also, the selection processes for individual cultural projects took much longer than planned and only a small part of the bid books could be considered. As a result, many individual artists in the independent sector were denied access to participation (see Mittag 2019: 201).

The lack of participation in the ECoC programme was also attested to by those responsible for RUHR.2010 in relation to the socially weaker sections of the population, which includes encounters between the different strata of the population: "The whole area of interculture has been very neglected," claims Jens Dirksen in the interview and further doubts that the migrant festival melez "worked" (Jens Dirksen 2016: in interview). Similarly, Jürgen Fischer concedes "that the Capital of Culture did not open up to the sometimes very difficult social situation in the municipalities. The Capital of Culture did not dare to actually address the real social problems. [...] I think the criticism that there was a lack of social integration in the programme is justified" (Jürgen Fischer 2016: in interview). Apparently, despite a "consistent low-price strategy",[20] the cultural offerings reached fewer of the so-called non-visitors[21] than had been intended by those responsible for the programme.

The "European dimension" is often at the back of the triad of ECoC criteria (sustainability, participation and a European dimension). As with the previous Capitals of Culture held in Germany, this was based on a lack of EU awareness among those responsible for the programme in RUHR.2010 (see the interviews with Jürgen Fischer and Hanns-Dietrich Schmidt, in Jacobsen 2009: 54 ff and more subtly formulated in the interviews with Gerald Baars, Jens Dirksen, Jürgen Fischer, Jürgen Mittag and Karl Schultheis 2016). Despite particular projects such as the extensive TWINS series, in which the European twin cities of the 53 cities of the Ruhr region contributed to the

---

20 See the evaluation report of RUHR.2010, which is very positive in its presentation and was partly co-authored by those responsible for the programme: Centre for Cultural Research/ICG Culturplan (2011): Mit Kultur zur Metropole? Evaluation of the Capital of Culture RUHR.2010, see link 25.

21 See various papers on non-visitor research by the Department of Cultural Policy at the University of Hildesheim, including Renz (2014, 2015) and Mandel & Renz (2010).

ECoC programme, "the European part [...] was certainly too weakly pronounced [...] minor and subordinate to the other accents of the programme" (Jürgen Mittag 2016: in interview). Cultural exchange and networking at RUHR.2010 were primarily focused on the region – thus the "European dimension" of the ECoC concept was pushed into the background.

RUHR.2010 also missed out on potential at the governance level, especially with regard to the legacy of the ECoC. According to Oliver Scheytt, this mainly concerns the cooperation of the "political class" in the Ruhr region:

> The city leaders and the local politics could have 'taken off' more strongly in order to use the unity in diversity for an overall appearance vis-à-vis the federal government and the state, especially since the citizens have developed a much stronger sense of community than was noticed and perceived by the city leaders (Oliver Scheytt 2018: in interview).

Last but not least, the promises made by those responsible for the programme with regard to a sustainable improvement of the cultural infrastructure in terms of mobility have not been fulfilled. Although shuttles and other special transport services were made available during the ECoC year, there was no lasting improvement in the frequency of public transport to make it easier to reach cultural sites. So, after the ECoC year, transport options in the less populated areas and between the cities of the Ruhr region in terms of accessibility to cultural venues still need to be improved.

6.5 Preliminary conclusion II: After-effects and needs after RUHR.2010

Even if its "trend towards eventisation" (Betz 2011: 333) cannot be denied, the ECoC RUHR.2010 is widely regarded as a high-profile and progressive example of the entire EU initiative (see Geilenbrügge & Eisermann 2014: 300). If one takes the large number and variety of projects, the diverse media coverage and the increase in tourist numbers as a yardstick, RUHR.2010 can be considered a success (see Mittag 2019: 200). For this reason, numerous other cities are examining which best-practice examples they can adapt from RUHR.2010. Thanks to the professional PR strategy aimed at an international audience, as well as the media networks of RUHR.2010 managing director and former WDR director Fritz Pleitgen, delegations from China and the Urals visited the programme managers to see how their own locations could benefit. The applicant cities for the next ECoC, which will take place in Germany again in 2025 for the first time since RUHR.2010, also took their cue from the last ECoC in this country. It is considered an example of great professionalism in the preparation phase, the organisational structures, the programme diversity and the marketing. Likewise, the motto of RUHR.2010 was adopted in modified form by later ECoCs (see Jürgen Mittag 2016: in interview).

Former RUHR.2010 employees also developed the passing on of experience into a business model, with varying degrees of success; they thus influence the still limited number of consultants who offer their services in the participating cities or even those cities which are interested in participating. In this way RUHR.2010 continues to influence the entire ECoC initiative to this day.

Like RUHR.2010, those cities that envisage a larger regional expansion for their programme face similar challenges of regional or cultural governance. How is it possible to ensure that all participating municipalities are equally committed to the implementation of the individual ECoC strategy within the scope of their capabilities? And if there is a political consensus, what does the functional sharing of joint funding look like? Diverse network building is also essential for the participating (candidate) cities: In which formats can the cultural institutions of one sector be brought together to plan series of joint events, for example? And an even more practical issue: How do visitors from regional city A get to regional city B to attend a cultural event if the local public transport system is not geared towards providing evening transport? The latter example makes it clear that the impulses suggested by the ECoC initiative go far beyond the will to reorient municipal cultural policy. Here the necessity of governance management in the multi-level system becomes clear, when "decisions between levels have to be coordinated" (Benz 2004: 127), i.e. the tasks, resources, duties and rights of different instances have to be coordinated.

Due to the many people and institutions involved, conflicts of interest are inevitable. If an attempt were made to please all parties, the individual profile of the ECoC concept would be lost. The monitoring by the EU Commission, which begins with the submission of the bid book, is not only intended to serve as an overall check of the formal criteria, but also to help clarify the governance structures. However, more and more participating cities are questioning whether the selective meetings with the selection panel or the "managing authority" actually support the increasingly complex ECoCs.

Already in the bidding process, the participating cities are exposed to the attentive observation of the actor networks from the participating sectors (culture, economy, tourism, etc.). The Europe-wide level of awareness, the extensive use of resources and the associated high expectations from the population contribute significantly to this. The bidding offices are thus under considerable pressure to succeed without having any systematic means of comparison.

Regrettably, it has not yet been possible to establish a central information and coordination point where ECoCs and candidate cities are at least provided with material to benefit from examples of success and good practice from the series. Of course, each city has to decide for itself, based on the conditions of its location, which type of governance will work for it. Nevertheless, it would be helpful if an archive of evaluation reports and documentation could be used to make the effects of an ECoC year (on municipal cultural policy, tourism and changes in cooperation and audience structures, etc.) usable for future ECoCs. The programme contents would therefore be less important in this mutual learning due to the very heterogeneous location conditions than an exchange about governance issues and the fundamental project management of an ECoC.

So far, this task has been taken over by individual networks or bodies, such as the informal network of the ECoC Family. The criticism here is that the meetings of the ECoC Family take place in camera and even neutral process observers are not allowed to participate. On the academic side, the University Network of the European Capitals of Culture (UNeECC, since 2006) and the ECoC LAB of the Department of Cultural Policy at the University of Hildesheim (2017–2021) have so far accompanied the further development of the initiative. The need for an information and qualification centre that promotes exchange between ECoCs and could perhaps stimulate joint concepts has now been brought to the attention of the EU Commission. Since the ECoC initiative has now been running for over 35 years, action by the EU Commission is long overdue.

7. Excursus: The role of urban and rural areas within the European Capital of Culture Initiative

In the run-up to RUHR.2010, the European Commission made it clear that the ECoC title could be awarded only to a city, not a region. For this reason, the official name "Essen for the Ruhr" was used for RUHR.2010 from then on. Brussels gave in about the name (see Mittag 2012: 68), but a programme for the whole region was nevertheless realised: in a way, each of the 53 cities of the Ruhr region was a Capital of Culture for a week – and Essen, as the official title holder, was the Capital of Culture for the whole of 2010.

Even then there was discussion: Does the spatial expansion dilute the concept of the ECoC? After all, the initiative in the early 1980s explicitly created the Capital of Culture as a "model of urban culture", "following the long tradition of a singular European urbanity, which architecturally brought the idea of Europe to the attention of its citizens and the world" (Schwencke 2010: 329).

The 1999 ECoC Decision therefore states in its recitals: "Throughout its history, Europe has been the site of exceptionally prolific and varied artistic activity; [...] urban life has played a major role in the growth and influence of the European cultures."[22] Thus, the intrinsic cultural logic of cities is to be seen as a condensate of holistic cultural-historical and social developments in Europe: "Cities are highly complex social and spatial entities. What inner logic do they follow? What gives them meaning, what makes them unique? What hidden structures underlie their actions?"[23] In this sense, the ECoCs are also a kind of laboratory for a future way of living together in the EU: "As a special urban entity, it is always the place where

---

22 Decision 1419/1999/EC of the European Parliament and of the Council of 25 May 1999, available at link 3.
23 See the project description of the LOEWE research project on the "Eigenlogik der Städte", available at link 18.

diverse new ideas develop and where its citizens feel at home. Thus the city is something like the guiding narrative of Europe" (Schwencke 2019: 57).

Over the years, however, the previously purely urban character of the ECoCs changed. Their former framework was therefore expanded not only in terms of the duration of the programme (from a summer festival at the beginning of the initiative to a multi-year cultural strategy) and the content, but also in terms of the geographical space in which it took place. In this respect, the RUHR.2010 programme for the first time clearly went beyond the EU's selection criteria. Since then, the subsequent ECoCs – even if this had not been envisaged as such until then – now also increasingly wanted to include the rural areas surrounding them in their agenda. EU legislation also expanded its guidelines in this respect: While the 1999 ECoC Decision in Art. 5 (see link 3) still states succinctly that "cities may choose to involve their surrounding region in their programme", the 2006 Decision (Recital 3, see link 6) recognises a greater effectiveness through regional expansion: "By enabling cities to involve their surrounding region [...] a wider public can be reached and the impact of the event can be amplified." The latter decision also aims to ensure the linkage of cultural policies between the city, regional and national levels: "The programme shall be consistent with any national cultural strategy or policy of the relevant Member State or, where applicable under a Member State's institutional arrangements, any regional cultural strategies [...]" (ibid., Art. 3,3).

This new relationship between cities and regions became more and more of an issue for the subsequent ECoCs, because it offered them many advantages:

- more inhabitants can be involved in the programme
- art in the rural areas can be a fruitful extension of urban cultural offerings
- the number of sights that attract tourists can be increased
- hidden treasures in the region can enrich the cultural landscape
- border regions can enrich the intercultural exchange of cultural institutions and organisations as well as the exchange between artists.

As for the above-mentioned concern that the urban idea of the ECoC might be weakened by the inclusion of regions, the current decision of 2014 (see link 5) directly refers to this and states: "The title should continue to be reserved for cities of all sizes, but cities should continue to include surrounding regions." This decision, which extends the provisional end of the initiative to 2033, makes direct reference to this and clarifies: "The title should continue to be reserved to cities, irrespective of their size, but in order to reach a wider public and amplify the impact, it should also be possible, as before, for the cities concerned to involve their surrounding area" (recital 12); in addition, "[w]here a candidate city involves its surrounding area, the application shall be made under the name of that city" (Art. 4,1).

The fact that this legal development of EU cultural policy has led to a strengthening of the importance of regions within the ECoC initiative is part of the overall regional policy of the EU. The multifaceted potential of rural areas has long been recognised and appropriate instruments of support[24] have been set up. The European Commission identified the most important goals and challenges in 2008 in its working paper "Regions 2020" (see link 4). Overall, regional policy, but also other EU policies, have the following advantages for rural development:

- *Economic advantages*: The objective is to strengthen the competitiveness and infrastructure of the regions. For reasons of cohesion, it is also a question of strengthening the weaker peripheral regions of the member states (e.g. by promoting and developing "Euroregions" or "European regions").
- *Political-administrative advantages*: In accordance with the principle of subsidiarity, regional administrations are considered to be more efficient because they have more expertise in their areas of responsibility and are closer to citizens. Since the 1980s it has also been postulated in the European Studies and Social Sciences that the nation states are too small to solve the big problems and too big to solve the small problems.[25] Article 198a of the Maastricht Treaty (1992) therefore

---

24 The main financial instruments of EU regional policy are: ERDF (European Regional Development Fund), ESF (European Social Fund), Cohesion Fund, EAFRD (European Agricultural Fund for Rural Development), JASPERS (Joint Assistance in Supporting Projects in European Regions), etc.

25 See Bell (1988: B3): "The common problem, I believe, is this: the nation-state is becoming too small for the big problems of life, and too big for the small problems of life. It is too small for the big problems because there are no effective international mechanisms to deal with such things as capital flows, commodity

established the Committee of the Regions (CoR) as a consultative body of the EU. But, beyond the EU, initiatives were also developed during this time which sought to give the regions in Europe a voice, such as the Assembly of European Regions (founded in 1985) or the Congress of Local and Regional Authorities of the Council of Europe (founded in 1994 as the successor institution of the Permanent Conference of Local and Regional Authorities of Europe).

– *Social and societal benefits*: Synergies can be achieved by developing common approaches to address Europe-wide challenges such as rural depopulation. This can go far beyond economic development: often, it is not the economic situation but rather the sense of belonging to a certain identity and perspective that decides where people want to live (an urban example is Berlin).

– *Cultural policy advantages*: Europe's cultural heritage does not end at the borders of nation states, but is rather to be found in the (also cross-border) regions. A counter-argument against EU cultural policy, especially in France, is that the regions of Europe are the true carriers of European culture, and not the EU.

The bid books of the candidate cities of the last years reflect the economic, social and cultural policy approaches of a Europe of Regions. The candidate cities Sønderborg (for the ECoC title of 2017) or Görlitz (for 2010) placed the common cultural area with their neighbouring countries in the foreground. This was an obvious move to fulfil the EU criterion of the *European dimension* – but in both cases this was not enough to convince the selection panel.

The city of Zittau in the border tripoint had similar interests with its bid. Like all the other candidate cities seeking the title of the next ECoC in Germany, it also wanted to involve the surrounding region. Other examples from the bidding process for the 2025 title are Hildesheim, which focused on a district with more than twice as many inhabitants as the city itself,

---

imbalances, the loss of jobs, and the several demographic tidal waves that will be developing in the next twenty years. It is too big for the small problems because the flow of power to a national political center means that the center becomes increasingly unresponsive to the variety and diversity of local needs. In short, there is a mismatch of scale."

and Nuremberg, which wanted to include the entire Nuremberg metropolitan region in its ECoC programme, which itself has a larger area than the federal state of Hesse.

Especially because the ECoC initiative has become so professionalised in recent years, there are various successful examples of ECoC programmes that do not stop at the city limits. These include the cultural walking trail around the city of San Sebastián (ECoC 2016), the programme and venues of the "Baroque Summer" in the Pilsen region (ECoC 2015) and the new museum network of the Midtjylland region (from the ECoC Aarhus 2017).

8. Marseille-Provence 2013

8.1 Historical, geopolitical and social conditions in Marseille and the surrounding area

The special conditions for the selection, preparation, implementation and sustainability of the ECoC programme in Marseille can only be explained by the history of the city. Thus, the central theme of the ECoC programme – the exchange with the cultures of the Mediterranean – was also the founding moment of the city. When Greek seafarers from Phocaea founded "Massalia" (its ancient name) around 600 BC by building the harbour, the city was already imprinted with a character that has remained to this day: Mediterranean-multicultural, cosmopolitan, and characterised by exchange with foreigners. Today, visitors to the city are amazed at how Marseille's inhabitants, from so many different cultural, social and religious backgrounds, can live together so peacefully. This can probably be explained by, among other things, the fact that the city has always facilitated diverse facets of identification for its inhabitants, even though the social divisions undoubtedly posed and continue to pose great challenges to them.

After its foundation, Massalia quickly grew into one of the richest and largest colonies of Greece in the Mediterranean. The river Rhône, as a natural trade route, guaranteed it economic and cultural influence in the northern and north-eastern areas. Even today, the foundation of Marseille by the Greek settlers plays a role in the city's relationship with its region, because the towns in the surrounding area are of Celtic-Ligurian or later Roman origin (such as Arles and Aix-en-Provence). These different founding moments still shape the mentalities of the inhabitants of the towns in the region and their rivalries and demarcations with each other.

Marseille was not integrated into the province of Narbonensis of the Roman Empire until 49 BC by Julius Caesar. The unification with France occurred centuries later (in 1481), after Marseille had previously been independent (since 1216 or 1218). Early on, these developments were associated with an esprit among the population that characterises the city to this day as particularly proud and independent. In 1792, the city sent 500 volunteer fighters to Paris to revolt against the Ancien Régime of the monarchy and support the French Revolution. Their blood-curdling battle song was later named the "Marseillaise" and declared the French national anthem in 1795.

In the 19th century, Marseille developed into the most important port of the French Empire. Both French colonisation in Africa and Indochina and the opening of the Suez Canal in 1869 contributed to this. In the course of industrialisation in the mid-19th century, a shift began in the city that, from today's perspective, is comparable to current gentrification processes. In the north of the city, homogeneous working-class neighbourhoods emerged near the new port of La Joliette. A similar second development occurred in the mid-20th century, when a large part of the industrial buildings (from the metal, chemical, food and "colonial goods" sectors) migrated to the outskirts of Marseille. To date, a cohesion policy has been neglected in order to compensate for the knock-on effects of the uneven economic and social development of the various districts. In the ECoC year, there were selective attempts to address grievances in this area, e.g. through an improved bus service to the northern districts (*quartiers nord*). However, this was withdrawn again after 2013.

After the Second World War and into the 1970s, the population of Marseille rose sharply due to the influx of inhabitants from France's former colonies in North Africa. After Algerian independence in 1962, tens of thousands of Algerian French who had to leave their country, so-called "*pieds noirs*", came to Marseille. They were in turn settled in the northern districts of the city.

From the 1970s onwards, the negative development of the city began, as a result of various factors: deindustrialisation in traditional sectors such as heavy industry and shipbuilding, uncontrolled and illegal immigration, increasing criminality of drug gangs, high youth unemployment and poverty, overtaxing of the local government with environmental pollution, and a growing volume of traffic. At the beginning of the 1970s, racist riots against Algerian immigrants increased, with up to a hundred deaths in 1973 alone. This development consolidated a different image of Marseille for the French and international public: as a dangerous, uncontrollable juggernaut, a kind of "failed city" that could no longer fulfil its basic functions.

This image of Marseille has been taken up artistically many times since then, e.g. in crime novels in which the beauty of the port city is contrasted with the murderous drug milieu. In the 1980s, young people from the socially deprived areas of Marseille developed French hip hop (for example, the band IAM), which today is a permanent fixture in the French musical culture alongside traditional chanson.

It was only in the 1990s that the city experienced a resurgence that continues to this day.[26] Marseille, which is closer to Algiers than to Paris,[27] was to be integrated more strongly into the French Republic, and at the same time its potential as a gateway to the Mediterranean was rediscovered. Various political decisions were made and measures taken to this end, which are briefly described below.

(a) The Euroméditerranée urban development programme

The implementation of the Euroméditerranée programme, "one of the most ambitious urban redevelopment projects in Europe", cannot be overlooked in today's cityscape (Woker 2015: 2). It was launched in 1989 on the initiative of the then mayor Robert Vigouroux and is still continuing today. Euroméditerranée aims to ensure that Marseille catches up with the development of other French cities and metropolitan regions. Priority is given to the policy areas of transport (mainly public transport), social affairs (including social housing), tourism and regional development, which were previously inappropriate for Marseilles as the second largest city in France.

Following the example of the redevelopment of Lyon's railway station district, the city quarters of La Belle de Mai, Saint-Charles, La Joliette, Arenc and the street Rue de la République are to be rebuilt and upgraded in the multi-billion project, which is jointly financed by the EU, the French state, the city of Marseille, and several regions and municipalities and

---

26 See Woker (2015): "Anyone who has eaten a bouillabaisse at the Vieux Port and has been chauffeured in a tourist train through the narrow streets to Notre-Dame de la Garde will want to confirm his own experience after his return that Marseille is no longer the drug hell and mafia stronghold that the city was considered to be two decades ago. In fact, the latest crime statistics show a convergence with other major French cities."

27 See the logo of the programme booklet: an upside-down map illustrates that Marseille is closer to Algiers than to Paris, and thus underlines Marseille's cultural proximity to and influence by the Mediterranean region.

private investors. In 1994, the *Etablissement Public d'Aménagement (EPA)* was founded for this purpose. In the district north of Place de la Joliette, hundreds of thousands of square metres of office space have already been created on a total area of over 3 km² – often in previously run-down residential neighbourhoods. Since the redevelopment of the district, more than 800 companies have settled there. To relieve the increasing traffic, the tram line T2 was installed on Rue de la République and the road tunnel *Tunnel de la Major* was opened in 2002. In 2005, the large passenger port for cruise ships opened in the La Joliette district.

The *Musée des Civilisations de l'Europe et de la Méditerranée (MuCEM)*, which opened in 2013, became the face of the ECoC MP2013, and subsequently *the* building of Marseille. Designed by the Algerian-born French architect Rudy Riciotti, it is now one of the three busiest tourist attractions in the city, along with the Old Port and the Stade Vélodrome. The adjacent *Villa Méditerranée* by Stefano Boeri, which towers horizontally over the water, was also opened in the ECoC year.

Directly at the harbour, a modern skyline is also being built along the street Les Quais d'Arenc. Already completed there are the *Tour CMA CGM* by Zaha Hadid (100m high), the *Tour Horizon* by Yves Lion (113m high), the *H99* by Jean-Baptiste Piétri (99m high), and *Le Balthazar* by Roland Carta (31m). The skyscraper *La Marseillaise* (135m high) by Jean Nouvel opened in 2018.

Roland Carta is also the architect of the port's former granary *Le Silo*, which opened in 2011 and now functions as an event centre similar to Hamburg's *Elbphilharmonie*. Its 2,000-seat concert hall hosts a diverse cultural programme ranging from ballet to comedy. The charm of the port industry has been preserved in several of the new or redesigned buildings. These include the *J1*, which once served as a hangar and was the central information point during the ECoC year.

(b) The Charter of Paris for a New Europe

This fundamental international agreement on the creation of a new peaceful order in Europe after the end of the Cold War was signed by 32 European countries, as well as the USA and Canada, at a special CSCE summit conference in 1990. Among other issues, it sets out the goal of increased cooperation with the countries bordering the Mediterranean Sea and was seen by Marseille as a prelude to economic and political cooperation in the Mediterranean region at the end of the 20th century (see link 27).

(c) The Euro-Mediterranean Partnership (EUROMED)

EUROMED was launched in 1995 at the Conference of Foreign Ministers of the EU and the Mediterranean partner countries. The partnership is also known as the *Barcelona Process* because the conference venue was Barcelona. Its objectives are to improve trade and other cooperation agreements between Europe and the neighbouring countries of the Mediterranean region, which the Council of Ministers designated as an "area of strategic importance" at the conference. Since 2004, there has been a permanent parliamentary supervisory body, the Euro-Mediterranean Parliamentary Assembly, and since then the *Barcelona Process* has also been complemented by instruments of the European Neighbourhood Policy (ENP). In 2008, the European Council decided to transform it into a *Union for the Mediterranean.* The goals of shared prosperity, peace and stabilisation, as well as a medium- to long-term transformation of the partner countries through economic and political cooperation, remained. It has not yet been possible to establish a Euro-Mediterranean Free Trade Area (*Deep and Comprehensive Free Trade Agreement – DCFTA*). Nevertheless, bilateral association agreements have been concluded and implemented between the European Communities and their member states and the Mediterranean partner countries, replacing the first generation agreements, i.e. the cooperation agreements concluded in the 1970s. Between 1998 and 2006, agreements were concluded with the Republic of Lebanon, the People's Democratic Republic of Algeria, the Arab Republic of Egypt, the Hashemite Kingdom of Jordan, the State of Israel, the Kingdom of Morocco, the Republic of Tunisia and the City of Tangier (see a summary of the legal acts of the Euro-Mediterranean Association Agreements in link 11).

(d) The extension of the high-speed line LGV Méditerranée (*ligne à grande vitesse méditerranée*)

Since 2001, the extension of the TGV line has enabled a journey time of only three hours between Marseille and Paris. It has also significantly improved traffic to the surrounding regions of Provence-Alpes-Côte d'Azur, Occitania and Auvergne-Rhône-Alpes.

For Marseille, too, the measures outlined for the Mediterranean region brought about medium- and long-term changes in its overall economic situation and helped Marseille to catch up on development. More concrete and immediate are the urban development measures, whose acceptance by the population, however, has not been trouble-free. Critics see the Euroméditerrannée project as a gentrification development which will systematically drive out the previous residents of the redevelopment areas. The association *Un Centre Ville Pour Tous* (*A city centre for everyone*, see link 28) campaigns for the residents who feel disadvantaged. The upgrading of the Rue de la République has stalled after extensive redevelopment. Although the city invested heavily there and also attracted companies by offering tax benefits, there is currently a lack of buyers for the prices of around 5,000 Euros per square metre. In other districts in the centre, such as Noailles, Le Panier, Chapitre or Belsunce, the square metre or rental prices are considerably lower. But this is inevitably linked to the fact that, despite their central location, these are very poor neighbourhoods, with unemployment of up to 30%. The proportion of people living below the poverty line is above average in Marseille. Between them and the wealthy upper class, who mostly live outside, i.e. mainly on the southern edge of the city, there is hardly any middle class. The *bobos* (short for "bourgeois bohemians": a type of well-off, environmentally conscious and individualistic city dweller) do leave their mark on the city, for example, through vegan restaurants, microbreweries and DIY designer shops in the Cours Julien neighbourhood, but they do not constitute a large group in the city population (see Megerle 2008, 2011 and 2017).

8.2 The ECoC as a driver of metropolitanisation and tourism promotion

Like RUHR.2010, MP 2013 is considered a particularly diverse and generally successful example of the ECoC initiative, with its programme covering a large geographical area. MP 2013 offered 400 projects, 60 exhibitions, and numerous concerts and artistic events in the various districts and cities of the region. The overall content of MP 2013 focused on the Mediterranean and had three thematic programme pillars: "Marseille-Provence welcoming the world", "Marseille-Provence Open Sky" and "Marseille-Provence, Land of Thousand Faces".

Together with its surroundings, Marseille implemented many of the EU's Capital of Culture requirements in a polymorphous way during the programme year, using the EU's initiative as a catalyst for further development and change. The most visible example of this is the redesign of the Old Port by renowned architects, first and foremost including the *MuCEM*, which has become Marseille's landmark. In addition to the urban development innovations, which also include the redesigned district of La Joliette, the art and culture scene has also experienced an upswing since the ECoC year. The former tobacco factory *La Friche La Belle de Mai* (*Friche* for short) has now developed into a large open cultural centre frequented by interested people from all over the city. Since the ECoC year, investments have been made here both in the expansion of the area and in its content, so that the *Friche* is now perceived as equal by artists, gallery owners and producers from Paris and has already received substantial international attention.

The criterion of fostering "the contribution of culture to the long-term development of cities in accordance with their respective strategies and priorities",[28] which only found its way into the decision underlying the ECoC series in 2014, was already taken into account in MP 2013. Evaluations such as that of the COM credit MP 2013 with a boost in urban development, with effects on cultural governance actors and on tourism: "For Marseille-Provence, the main legacy effects were increased recognition as a cultural and tourism destination and strengthened networking between cultural operators and policymakers across the territory and internationally".[29]

---

28 Decision No 445/2014/EU of the European Parliament and of the Council of 16 April 2014 establishing a Union action for the European Capitals of Culture for the years 2020 to 2033 and repealing Decision No 1622/2006/EC. See link 5.

29 European Commission: Ex post evaluation of the 2013 European Capitals of Culture (p. 3). See link 12.

However, the question arises as to how well-founded the label "cultural tourism location" really is with regard to Marseille. It is true that tourism marketing research has meanwhile discovered that visitors to Marseille regard the MuCEM as one of the three most important tourist magnets (together with the Old Port and the Stade Vélodrome, the home stadium of the Olympique Marseille football club, see above). But it is questionable whether the MuCEM is visited by the large number of (cruise) tourists for its changing exhibitions or to take selfies on the roof garden in front of the fantastic harbour panorama. Even though there is no study on this yet, it seems that many of the MuCEM's visitors spend most of their time on the picturesque roof terrace and not in the exhibition rooms.

Essential to understanding the achievements of MP 2013 is the fact that the city's application to become an ECoC did not come from the cultural scene or administration, but from the private sector. The Chambre de Commerce et d'Industrie (CCI) and the "Top 20", the city's 20 leading industrial companies, exerted political pressure to develop metropolitanisation in Marseille and the surrounding area, as had already been done successfully in other cities in France. In the mid-2000s, following the example of the last ECoC in France (Lille 2004), the idea was born to promote metropolitan development through a major cultural event. MP 2013 thus became a pioneering project, both for the city of Marseille and for the surrounding 93 municipalities of the Département Bouches-du-Rhônes.

Overall, MP 2013 was thus clearly driven by the private sector, and consequently the president of the CCI, Jacques Pfister, "an ardent advocate of metropolitan development" (Ulrich Fuchs 2016: in interview), was appointed president of the Capital of Culture. In 2009, the marketing director of Euroméditerranée also moved to MP 2013. His move was intended to build even more intensive relationships in the private sector in order to acquire sponsorship (see Liehr 2013: 285). In addition to the political representatives of the city of Marseille, the Conseil Général (Département) and the Conseil Régional (Région) and the cultural representatives, the supervisory board of MP 2013 also included representatives of the Grand Port Maritime, the airport and other Euroméditerranée employees. These personalities alone make the economic orientation of MP 2013 very clear, from the first plans for the bid.

8.3 Governance, intrinsic logic and obstinacy of the Marseillais in the ECoC context

After Marseille was awarded the ECoC title, many disagreements arose between the cities involved. MP 2013 was seen as a "Trojan horse" that was only put forward in order to obtain more funds for metropolitanisation from the French Republic, but also from other levels.[30] Overall, cultural governance between the city and the region was difficult. In comparison to the governance conditions at RUHR.2010, there was not only "parochial thinking" (see chapter 6.1) at MP 2013: a "provincial cultural clique" (Liehr 2013: 286) was very pronounced in all the participating municipalities of MP 2013, as a result of the history described. In both the author's background discussions and expert interviews, statements that illustrated the difficult conditions for cooperation in MP 2013 were frequently made:

> The people from the region don't like Marseille that much. [...] If you live in Aix and someone wants to bring you here, everyone thinks you are stupid. It still shows today: Marseille is a Greek-founded city and incredibly open and friendly, and everyone here is an immigrant anyway, no matter how many generations they've been here. And Aix is a Roman-founded city that has never demolished its walls, and anyone who even comes near the city walls is a potential enemy. This is the attitude of Marseille and Aix to this day (Pia Leydolt-Fuchs 2016: in interview).

Or: "Marseille [is] a particular example of how people do not work together" (Antonia Blau 2016: in interview).

These suboptimal basic conditions for joint cooperation in the sense of the governance approach meant that municipalities such as Aix-en-Provence were only involved to a small extent in MP 2013 and thus missed out on the possibilities such as those used by Marseille for its urban development.

Ulrich Fuchs, who was the deputy artistic director and programme director of both Linz 2009 and MP 2013, assessed the governance conditions in Marseille as difficult, also in comparison to Linz 2009:

> The governance of this project [MP 2013] was extremely complicated and cannot be compared with my experience in Linz, for example. Although in Linz the city

---

30 This was expressed by several mayors of the Provence-Alpes-Côte d'Azur (PACA) region and summarised in film by the Marseille documentary filmmaker Nicolas Burlaud: see link 31.

and the surrounding area were also governed by different parties, the common will for the Capital of Culture brought about a largely harmonious cooperation. Here [in Marseille], however, the ECoC project was partly taken hostage, à la: "If we can't push through this or that political project vis-à-vis Marseille, we'll pull out of MP 2013" – even though one had nothing whatsoever to do with the other (Ulrich Fuchs 2016: in interview).

Against this background, the interests of those responsible for shaping a joint programme for the city and the region were not only economically oriented, but also had a political and not least administrative intention. The previous lack of willingness to cooperate on the part of the municipal and regional actors had led to various duplicate structures, as the following example illustrates: Before MP 2013, there were various street theatre and circus companies, but they had never worked together. For the ECoC year, the region's potential in this field was recognised and the artists were brought together. This resulted in the *Biennale Internationale des Arts du Cirque (BIAC)* (see link 32), which still exists today and has become a prime example of the successful sustainability of an ECoC programme element.

But apart from positive individual examples like this one, MP 2013 did not succeed in sustainably improving the willingness of the actors to cooperate in the field of cultural policy: "Je pense qu'on a insuffisamment développé les logiques de réseau. Le but était de fédérer les acteurs. Cela a marché un peu dans certains secteurs, mais pas tant que ça."[31] (Jean-François Chougnet 2016: in interview). This is how the general director of MP 2013 summed it up critically in retrospect, thus conceding that the ECoC programme did not bring about any fundamental improvement in cultural governance in the city and region. Gilles Suzanne, Marseillais and academic process observer, finds even clearer words for what he sees as a failed attempt to make the cultural policy structures of Marseille and the region more permeable to each other: "Mais qu'est-ce que ça a installé comme dynamique entre ville et campagne? Cela est un grand néant. Qu'est-ce qu'on en retire?

---

31 "I think we did not pay enough attention to the logic of the networks. The goal was to unite the actors. That worked in some areas, but not really."

Rien, franchement rien. Tous les rapports ont révélé l'évidence: cela n'a pas fonctionné"[32] (Gilles Suzanne 2016: in interview).

8.4 Capital of Culture – by all and for all?[33]

With regard to the criterion of participation, MP 2013 revealed one of its greatest shortcomings. It overlooked *the* art form of Marseille, which artistically reflects the city's intrinsic cultural logic like no other: the rap music. "Marseille is not as famous for any other cultural asset as it is for rap music; the best rappers in France have been coming here for years. But in the year of the ECoC, their music hardly plays a role" (Bopp 2015). While other French cities, apart from Paris, have particular artistic genres – such as the theatre festival for Avignon, the film festival for Cannes, the Biennale for Lyon or the opera for Aix-en-Provence – Marseille's most authentic and successful art form is rap. The genre would have lent itself to the ECoC as a form of expression of the city's intrinsic cultural logic – after all, rap is all about *street credibility*, i.e. authenticity and credibility when it tells of life on the street. Because there are not many other cultural assets that have emerged in Marseille and achieved fame beyond the city's borders, rap culture was advertised in the bid book for MP 2013. Then, in the ECoC programme, there was only one concert series that brought together rappers from the Mediterranean, but it did not receive much attention.

In the face of this omission, rap gave a voice to the very opponents of MP 2013. Keny Arkana, the most famous female rapper in Marseille, gained a lot of attention and support with her song "Capitale de la Rupture" (instead of "Capitale de la Culture"). Her music video for this song, in which she denounces the ECoC project for being geared only towards gentrification ("Nos rues se remplissent de tristesse – un genre d'Apartheid se dessine"[34]), has already been viewed over 3 million times on YouTube.

---

32 "But what dynamic was achieved between the city and the countryside? It was a big nothing. What do we learn from that? Nothing, frankly, nothing. All the reports show this obviousness: it didn't work."

33 This claim is very often made in the ECoC discourse at conferences and discussion events: see Kolb & Scheytt (2010).

34 "Our streets are filled with sadness – a kind of apartheid is taking shape."

A showcase project of MP 2013 in terms of district culture is the above-mentioned cultural complex *La Friche de la Belle de Mai* near Marseille's main railway station Saint-Charles. It was significantly expanded for the ECoC year and now combines, among other things, offices for the creative industries, various cultural production facilities, event spaces and a socio-cultural centre under its roof. In the avant-garde, it has become an established location, as shown. But did the place really become a district centre in which the surrounding population of the working-class Belle de Mai district participates, as it was outlined in the ECoC bid book? "Ce que je ne veux pas c'est ghettoïser ce lieu. Il faut ouvrir ce lieu à l'ensemble du territoire"[35] (Alain Arnaudet 2016: in interview) insists the director of the *Friche*, Alain Arnaudet. But, on the other hand, the *Friche* is also perceived by local residents as elitist, as "an eminent cultural pole with workshops and event spaces in the neighbouring working-class Belle de Mai district, for whose residents it seems like a quaint foreign body" (Liehr 2013: 285). Antonia Blau, who worked in the *Friche* during the ECoC year and afterwards, has a similar opinion:

> While the *Friche* is successful overall, it is nevertheless also a self-contained microcosm. From the outside, the *Friche* is often perceived as not very accessible. Even if it looks so idyllic, with the many underprivileged children playing basketball in the courtyard and the idea being: "At some point you'll also take them to the exhibition ..." – but that doesn't hide the fact that the events tend to be attended by a specific group. One exception is the *Toit Terrasses* series, where the DJs play on the roof terrace, which is a huge success for everyone (Antonia Blau 2016: in interview).

Although the cultural offering of the *Friche* is undoubtedly very broadly oriented, there seems to be a danger that the site will only be frequented by the usual suspects. Here, as with other ECoC elements, it is evidently necessary to examine the cultural policy concepts with which MP 2013 was launched to ensure that the reality is closer to the participatory claims in particular.

Similar to the ECOC Leeuwarden 2018, there was already dissatisfaction among local artists with MP 2013 regarding their level of involvement in the programme (see chapter 6.4). The provocative statement from the

---

35 "What I don't want is to ghettoise this place. You have to open this place up to the ensemble of the area (district)."

CCI – "On ne peut pas jouer la Champions League avec une équipe de division d'honneur"[36] (Maisetti 2013: 60) – further fuelled this conflict. The fact that a supposedly disproportionate amount of money had been spent on internationally recognised artists enraged the local cultural scene. In background discussions it became clear that they would have liked more of the ECoC budget to have been used for smaller local projects. The costs of the opening ceremony of the ECoC year (2.9 million Euros) were also criticised in this context as being too high.

8.5 Preliminary conclusion I: Sustainable transformation and cultural policy perspective

Cultural sustainability – initiated by the ECoC programme – which ideally leads to a sustainable transformation of the location can only happen if there are permanently adequate places where culture can take place. Many ECoCs throughout the history of the initiative therefore made a new or renovated cultural site the symbol of their ECoC concept, such as RUHR.2010 (Zeche Zollverein), Pilsen 2015 (new building of the City Theatre) or Aarhus 2017 (Dokk1). In the case of Weimar (1999), the structural part of the ECoC programme was essentially the only thing that remained sustainable. That the transformation of parts of Marseille's cityscape is sustainable cannot be denied: "As ephemeral as the spectacles of the Capital of Culture year may be, the event nevertheless leaves behind [in Marseille] a considerable new infrastructure that guarantees a wealth of further events in the future" (Liehr 2013: 286). For without the ECoC title, the MuCEM and the Villa Méditerranée would certainly not have been built, and the Old Port would not have been converted into a pedestrian zone either – at least not at that time.

What will remain, then, is access to the sea, previously blocked by a motorway and car parks, and the reclaiming of public space in the harbour area, which is now also enriched with several new cultural facilities. However, it should be noted that the new cultural facilities and offers are concentrated in the inner-city area. Despite investments in the hundreds of millions – 160 million Euros alone were made available for the construction

---

36 "You can't play Champions League with amateurs."

of the MuCEM – not a single new permanent cultural institution has been built in the northern arrondissements, the socially disadvantaged *quartiers nord*. Furthermore, there are still almost no municipal cultural centres there, and in the nationwide distribution of libraries, the northern districts even rank last in France with a quota of 400 m² for 100,000 inhabitants, including the rural peripheral areas (see Megerle 2017: 362).

As far as the accessibility of the newly created or already existing cultural sites is concerned, MP 2013 did not achieve any sustainable improvement in 2013. Mobility and transport accessibility remain challenging in Marseille. If you want to travel to Marseille, you have to deal with either the very frequently congested motorway or poor railway connections. In Marseille itself, there are still only two metro lines, and some parts of the city are even difficult to reach by bus. Although a third of Marseille's urban population lives in the *quartiers nord* (about 300,000 inhabitants), no metro or tram runs there: "It's like a crack is going through the city" (Liehr 2013: 287). During the ECoC year, the bus service was improved at certain points, but this could not be maintained permanently. This means that residents from the poorer neighbourhoods on the outskirts continue to find it difficult to participate in cultural life in the city centre, as was the case before 2013.

The cultural policy concept of the ECoC bid, according to which a programme was to be developed for the whole population, has at least partially worked out. In fact, those responsible for the new or expanded cultural venues for the ECoC year describe their audience as pleasingly heterogeneous.[37] Apparently, it has been possible to make offers low-threshold, to which the location of the cultural institutions at the harbour also contributes in some cases. The cultural policy approach of MP 2013, to have as many of the events as possible take place "for free and outdoors",[38] was also so successful that it was taken up by later ECoCs and candidate cities as a principle and guarantee of success for participation.[39]

---

37 E.g. Jean-François Chougnet (MuCEM) or Alain Arnaudet (La Friche La Belle de Mai) in interviews (2016).

38 Ulrich Fuchs: "Umsonst und im Freien erleichtert den Zugang zur Kultur", in: Prante 2015: 11.

39 For example, in Pafos 2017 or in the candidate city Nuremberg (as ECoC 2025).

The return of the Goethe Institut to Marseille can be seen as "institutional sustainability". It was (re)established there as part of the expansion and reorientation of the *Friche* and still exists today, albeit on a small scale. This can be subsumed under the ECoC criterion of the *European dimension* and at the same time underpins the intercultural claim of the *Friche*.

The COM's evaluation report provides a significant point of criticism with regard to sustainable culture-based change in Marseille. It criticises the lack of stable structures that allow for solid sustainability planning and the long-term impact of MP 2013: "In the absence of formal legacy structures or cultural competencies for Marseille-Provence Métropole, this is most likely to be done on an informal basis in future."[40] The goal of deepening cooperation in the Marseille-Provence region must therefore henceforth be pursued beyond the regular municipal political channels. In order to ensure the sustainability of what was achieved, a number of partner organisations should have been involved at an early stage and adequate resources should have been available (ibid.). In fact, no successor organisation or other legacy structures were created, as was the case with RUHR.2010 and also Lille 2004. The latter ECoC founded the organisation *Lille 3000* after the programme year, which has organised an international cultural event every two years since 2004. This builds on Lille's ECoC experience and is financed half publicly and half privately. But such self-commitments by the city were absent in Marseille.

When the final accounting of the 2013 MP budget was completed, it was discovered that there was a surplus of almost one million Euros. The city and the region then agreed to each contribute the same amount for the next major cultural event: the six-and-a-half-month cultural festival "MP 2018 – Quel Amour" (see link 33), which took place five years after the ECoC year and again in the city and the region. All in all, Marseille had the confidence to win further awards after the successful ECoC bid, which in turn also convinced the corresponding selection panels: for example, the title "European Capital of Sport", which was awarded to Marseille for 2017, and MANIFESTA (European Biennial of Contemporary Art), which took place in Marseille in 2020 despite the COVID-19 pandemic.

---

40 Ex-post Evaluation of the 2013 European Capitals of Culture, p. 74: see link 12.

8.6 Preliminary conclusion II: Quo vadis transformation?

As shown above, the influence of the Euroméditerranée urban development strategy cannot be ignored in the overall concept of MP 2013. Consequently, it would be inaccurate to attribute to Marseille a purely culturally and politically motivated transformation brought about by the ECoC programme. Rather, MP 2013 should be seen as an element of a comprehensive urban development process that is clearly perceptible to the population and that is far from being completed: "The bid, but above all the eventual selection of Marseille as Capital of Culture 2013 was an almost inevitable consequence of the comprehensive urban renewal process of recent years." (Megerle 2017: 349).

By means of the Capital of Culture programme and also afterwards, Marseille's local politics – similar to RUHR.2010 – was and is interested in achieving an improved city branding through the initiated transformation process and in further cultivating Marseille's new image as a cultural location. But there is also local criticism of the rapid changes. Parts of the population feel excluded or even disconnected and criticise the commercialisation and gentrification of traditional residential areas. In addition, many have criticised the fact that the cultural offerings are primarily geared towards cruise tourists instead of residents; Günter Lier quotes a cultural officer from the local government: "Culture is supposed to attract executives and tourists to Marseille." (Liehr 2013: 284). Jacques Pfister, who was president of both the CCI and MP 2013, put it somewhat more elaborately: "A strong cultural attractiveness on an international level is increasingly seen as a major advantage in the competitive battle that European metropolises are waging" (ibid.).

The most visible point of conflict in the transformation of the city is the conversion of Rue de la République, which connects the cruise port with the Old Port, described in chapter 8.1. On the one hand, the street was supposed to represent a business quarter in which thousands of square metres of office space were to be built. On the other hand, it was planned as a shopping zone through which tourists coming from the cruise port could be channelled directly to the Old Port. The influence of a municipal cultural policy on this newly designed district is not discernible here.

It remains to be seen to what extent urban development in the sense of Euroméditerranée will progress or how initiatives such as *Un Centre Ville pour Tous* will help shape the transformation process, because the changes have been causing social tensions for some time, especially in the city centre:

In particular, population groups that were already precarious beforehand were not only mostly unable to draw any benefit from the urban renewal processes and the actions within the framework of the Capital of Culture, but on the contrary even experienced sometimes clearly negative effects (Megerle 2017: 343f).

The ECoC is thus viewed more critically by the population in retrospect, compared to during the programme year.

The accusation of the "Trojan Horse", according to which MP 2013 was supposed to conceal or gloss over gentrification developments, is gaining new momentum through the protests of the *Mouvement des Gilets Jaunes (Yellow Vest Movement)*. From the point of view of opponents of gentrification, the collapse of two dilapidated houses on 5 November 2018 in the city centre with eight fatalities, reminiscent of the fire in London's Grenfell Tower on 14 June 2017 with 72 victims, also demonstrates that local politics is focusing too much on urban upgrading without taking socially weaker sections of the population into consideration.

9. Conclusion/Theses

9.1 Concluding remarks

The quotation from Richard von Weizsäcker that precedes this work has a sequel:

> Politics is always also cultural politics, because politics influences the conditions of people's lives and actions [...]. Conversely, every change in the cultural state of mind sooner or later affects politics. Time and again, political consciousness turns out to be the cultural consciousness of a concrete political decision-making situation (Weizsäcker 1987: 22).

In the field of investigation of this study, the interaction between politics and culture described by Richard von Weizsäcker became clear. The "cultural state of mind" was interpreted as the "intrinsic logic of cities" and specified as the "intrinsic cultural logic" of the cultural sector of cities. On the breeding ground of their intrinsic logics, a will for transformation emerged in the two cities analysed, which was set in motion in different governance settings. It became apparent that the ECoC created effective potential both in the Ruhr area and in Marseille-Provence to stimulate, support and advance the complex and long-term transformation processes in the cities. The motto chosen by RUHR.2010, "Change through Culture – Culture through Change", can in this respect be applied in gradations to all ECoCs – not least because the "change agenda" must be explicitly included in the bid book for the ECoC title. In the case of the ECoC, the same EU instrument with the same legal requirements is implemented differently in each location. Against the background of the actors' intrinsic cultural logic and interests, different (power) structures of governance are established and different decisions are made.

The analysis of the two ECoCs, RUHR.2010 and MP 2013, reveals that the locations used the EU designation for different functions. In the case of RUHR.2010, the ECoC year followed a process of establishing a regional understanding that began about 20 years previously. In the two preceding decades, the RVR (or previously: KVR) had already ensured that the Ruhr region developed solid forms of cooperation in several policy areas, such as public transport or road planning. The ECoC, as a kind of ideological superstructure, only took place ex post.

With MP 2013, on the other hand, it was the other way round: the initial spark for metropolitanisation only came from the ECoC. MP 2013 thus took place at the beginning of the fundamental transformation process for the city, which was only initiated after a long standstill in development. This can be seen in the vehement revolt of the opponents of gentrification, who not without reason point to the dark sides of urban renewal. In no other location of the initiative has there been comparable resistance to the ECoC programme so far.

9.2 Gains and future tasks of RUHR.2010 and MP 2013

A successful sustainability of the ECoC concepts is expressed, for example, in the permanently increased visibility of an ECoC. The repositioning of the two cities, the award by the EU, the presence of the extensive cultural programme in a relatively large area and, last but not least, the multi-million investments provided the media with numerous occasions for reporting that were able to bring the new self-image of the two locations to the world. Conversely, the media take-up of the Capital of Culture, especially in the region, was also important for internal perceptions in order to maintain interest in the ECoC's activities during the programme year. The "Capital of Culture as a giant marketing machine" (Hildesheimer Zeitung, 21 March 2017) thus had an effect not only externally but also internally.

In both locations, the visibility of their urban and regional profiles has measurably improved as a result of the ECOC programme. This visibility has the potential to improve the citizens' relationship with their city. The improved urban self-awareness, intangible but still perceptible, was formulated by Alain Arnaudet with regard to Marseille as follows:

> Je pense que ça a donné une nouvelle image de Marseille aux Marseillais et au monde également. «Notre ville est chouette, belle, originale, créative». Et je le vois également à l'extérieur [...]: «Quelle belle ville attractive, on adore Marseille». Je ne pense pas qu'on nous aurait dit ça il y a cinq ans.[41] (Alain Arnaudet 2016: in interview)

---

41 "I think it [the Capital of Culture programme] has given a new image of Marseille to Marseillians and the world alike. 'Our city is cool, beautiful, real

Jens Dirksen summed RUHR.2010 up as follows:

> Through the ECoC, the Ruhr region has gained the experience of being a sleeping giant, so to speak [...] and it has also gained self-confidence. [...] And the tourism thing has also worked. That is remarkable. Today you still hear the clichés that briquettes fly through the air here, and every second person who comes here says: Oh, but it's green here! We can no longer hear that here. It implies the expectation that everything here is actually black. The sky over the Ruhr has been blue for 20 years, not least because industry has gone bust. The Capital of Culture has now corrected that (Jens Dirksen 2016: in interview).

Through the self-assurance and responsibilisation, i.e. an increased feeling of responsibility of the citizens with their city, both RUHR.2010 and MP 2013 became a laboratory for an examination of challenges that affect many cities and regions in Europe, such as structural change in the region. The identity discourse initiated by the ECoC programme is thus the strongest feature of a "European dimension", which is in fact not defined anywhere in EU legislation. In this sense, it is therefore not true that the "European dimension" is the least pronounced component of the central ECoC criteria (sustainability, participation and European dimension) in the evaluations of the two ECoCs.

Even though RUHR.2010 and MP 2013 exceeded the legal requirements of their time as ECoCs, as presented in this analysis, challenges remain in both locations that will continue to play a role in cultural policy discourse in the future. As the study shows, the cultural sector in both the Ruhr and Marseille-Provence is in permanent danger of being instrumentalised by other economic and political players. In order to protect against such misappropriation, the public cultural mandate must be continually reflected and adapted to the lifeworld conditions of the present. This makes long-term strategic cultural development planning within the framework of urban development concepts indispensable. This is the only way to protect the diversity and intrinsic value of cultural life, which do not require external legitimation through higher-level goals, from the grasp of other interests.

---

and creative'. And I see that from the outside as well [...]: 'What an attractive city, we admire Marseille.' I don't think we would have been told that five years ago."

In this context, it also remains a permanent task to direct access to culture not only towards an academic elite: "In order to do justice to the significance of art and culture for the individual and society, a cultural policy is needed that particularly advances the process of cultural participation." (Schneider 2011: 13). For cultural policy is still far from a "culture for all", the motto of the New Cultural Policy of the 1970s formulated by Hilmar Hoffmann,[42] which MP 2013 and RUHR.2010 also revealed. Here it became apparent that "culture for all" is less the result of progressive cultural policy than it is the result of the influence of economic interests. A cultural oversupply that currently exists from profit-oriented, private-sector sides (but also through civil society, for example, on YouTube, the "largest educational and cultural platform in the world", see Renner 2016: 371) reveals that genuine cultural policy mandates have now been taken over by actors other than politics. As a result, cultural policy has been released from its "compulsory tasks" and is ultimately free to define new tasks itself. This opens up new possibilities for innovative cooperation, i.e. for new governance structures, both in terms of content and operations. "European Capitals of Culture can serve as an experimental area to try out new structuring and participation possibilities in this sense", which give new impetus to municipal and regional cultural policy. Of greater importance than the ECoC year itself is a sustained dialogue with the population about the meaning, function and responsibility of cultural policy decisions.

9.3 Theses

In accordance with the triadic analysis procedure of the Grounded Theory, conclusions were developed that represent an answer to the research questions formulated in chapter 1.4. With the help of the qualitative-explorative research design of the multimethod research presented in chapter 5.2, the following theses were developed:

T1: There is no finalité for cultural governance.
The bilingual phrase "There is not finalité" is often used in European studies discourse with reference to the progress of European integration and points

---

42 See Hoffmann (1981).

to its lack of finality.[43] It originates (slightly modified linguistically) from the *Reflection Group on the Spiritual and Cultural Dimension of Europe* under former EU Commission President Romano Prodi, which published the results of its work in 2004:

> There is no essence of Europe, no fixed list of European values. There is no "finality" to the process of European integration. Europe is a project of the future. With every decision, not only in its zone of peace, its institutions, its political, economic and social order, but also its very identity and self-determination are opened for questioning and debate (Biedenkopf et al. 2004: 12).

In relation to the ECoC initiative, thesis T1 means that cultural governance cannot reach a terminated saturation in the implementation of cultural policy goals. With regard to the current EU decision on ECoCs of 2014,[44] this means that the urban development strategy called for therein, which covers at least ten years, requires a permanent dynamic of cultural governance. The strategic goals of the EU must also be critically reflected upon with regard to the intrinsic value of culture. After all, in the latest EU strategic documents on cultural policy – the COM's *New European Agenda for Culture* (May 2018)[45] and the Council's *Work Plan for Culture* 2019–2022 (December 2018)[46] – culture is seen primarily as a resource to advance social cohesion, economic growth and the strengthening of external relations. The research question F2 on the cultural policy perspectives for the further development of the ECoC initiative remains relevant against this background insofar as the research results presented here provide orientation on the one hand, but, on the other hand, there is no "finalité" to be able to answer it conclusively.

---

43 For example: "Indeed, the beauty and sui generis character of the EU lies precisely in the fact that *there is no finalité*, no end goal to the integration project" (Gagnon et al. 2015: 13).
44 See link 5.
45 See link 14.
46 See link 16.

T2: The cities' own cultural logic is at odds with the "travelling circus" of the European Capital of Culture.
Due to the local specifics of cities in the cultural sector, adapting ECoC concepts that have been successful in another location is only useful to a limited extent, if at all (see chapter 4.2): "Transferability of the cultural policy and the shaping of local cultural spaces of a particular city to other cities is not possible due to the respective contexts, conditions and possibilities" (Könneke 2011: 114). The intrinsic cultural logic should therefore be included as a cultural policy concept in the ECoC discourse, because it has been established that it is precisely the richness of innovation in dealing with the local-specific challenges of cities that constitutes the essence of the ECoC initiative.

The discourse showed that the approach of the intrinsic cultural logic of cities can help to analyse the cultural specifics of a location in the context of an ECoC bid – the earlier in the process the better. It can also help the smaller candidate cities, in particular, as they have less capacity to prepare the bidding process, in looking for their self-location within the ECoC initiative.

Research question F1 on the cultural policy dispositions and conceptions of the two ECoCs, which was explored in chapters 6 and 8, should therefore be the starting point for all candidate cities in their bidding activities in the sense of the theory of the intrinsic cultural logic of cities as a basic requirement.

T3: The ECoC initiative represents an example of progressive urban development policy that can be a model for all cities.
When investigating research question F3 on the contribution of the ECoC concepts to cultural policy transformation processes in both selected locations (chapters 6 and 8), it became clear that a reliable cultural development strategy is an indispensable basic requirement for cities in implementing their ECoC programmes. The cultural development strategy is in any case one of the six binding selection criteria of the EU legal basis. Since the topic of cultural development planning has played a prominent role in the general cultural policy discourse for several years, the experiences of the Capitals of Culture can also be profitable for other cities not participating in the initiative, because the catalogue of questions that must be

addressed in the bid book is "a prime example of guidance for strategic thinking and action in cultural policy" (Oliver Scheytt 2018: in interview).

All cities can benefit from the professionalisation of ECoCs, especially with regard to the structures of cultural governance to be established at the intersectoral political-administrative, institutional and artistic levels. What needs to be clarified is what kind of knowledge and experienced management is needed at the specific locations in order to exchange and further develop expertise.

9.4 Summary and outlook

While the ECoC initiative initially had only a colourful festival character in its early days, over the years it has developed into an influential series that enabled the participating cities to meet their individual challenges in a new way through a cultural policy and interdisciplinary strategy. In fact, the ECoC initiative thus even exceeds the expectations of the European Commission, which acknowledges that even without its intervention, the initiative developed further than it would have previously thought possible (see Jacobsen 2017: 14). As this study has shown, the ECoC functioned as an essential component of comprehensive urban development processes, especially at the RUHR.2010 and MP 2013 sites. While the successes of the two ECoCs in the three categories of sustainability, participation and European dimension varied, their overall achievements have become a benchmark for cities participating in the initiative in the future.

Its successful operation in the networks of the European multi-level system and its creative ability to form coalitions of interests are also exemplary for other EU policy fields. For the ECoC initiative as by far the greatest instrument of success of European cultural policy, there is, as for many other EU initiatives, no precise predetermined roadmap: "Europeanisation 'takes place', 'acts' virtually in institutionalised improvisation" (Beck & Grande 2004: 65).

But despite all the colourful facets of the ECoC – the promotion of the creative industries, the new networks, the multifaceted cultural programme, the structural renovations of cultural sites, etc. – one should not underestimate what the initiative is also able to achieve ideally. For through the reflection of the lifeworld inherent in art and culture, it can help people to find opportunities for identification and to counteract the uncertainties associated with processes of change: "Artists could also be 'watch dogs', in the sense of being agents of changes through identifying and perceiving the inhibitors change" (Schneider 2016: 4 f). In the Europe of the present, in which, as was explained in chapter 1.1, the community no longer represents a trustworthy basis for its future for many citizens, the ECoC programme is an important element in making people aware of the common value foundations of Europe. Only in this way can the EU remain the "united in diversity" peace project that it presents itself as in the context of the ECoC initiative.

Bibliography

Arnaudet, Alain (2016): Expert interview conducted by Kristina Marion Jacobsen on 1 November 2016 in Marseille

Baars, Gerald (2016): Expert interview conducted by Kristina Marion Jacobsen on 18 February 2016 in Dortmund

Baier, Nikolaj & Scheytt, Oliver (2010): "Kulturhauptstadt", in: Lewinski Reuter, Verena; Lüddemann, Stefan: Glossar Kulturmanagement, Wiesbaden, pp. 150–159

Bärwolff, Theresa & Schneider, Wolfgang (2019): Vorwort, in: Schneider, Wolfgang; Butel, Yannick; Bärwolff, Theresa & Suzanne, Gilles (eds.): Dispositive der Transformation. Kulturelle Praktiken und künstlerische Prozesse, Hildesheim, pp. 15–20

Beck, Ulrich & Grande, Edgar (2004): Das kosmopolitische Europa. Gesellschaft und Politik in der Zweiten Moderne, Frankfurt

Bell, Daniel (1988): Previewing Planet Earth in 2013, in: The Washington Post, 3 January 1988

Betz, G. (2011): Das Ruhrgebiet – europäische Stadt im Werden? Strukturwandel und Governance durch die "Kulturhauptstadt Europas RUHR.2010", in: Frey, Oliver & Koch, Florian (eds.): Die Zukunft der Europäischen Stadt. Stadtpolitik, Stadtplanung und Stadtgesellschaft im Wandel, Wiesbaden, pp. 324–342

Biedenkopf, Kurt; Geremek, Bronislaw & Michalski, Krzysztof (2004): The spiritual and cultural dimensions of Europe: Concluding Remarks, Wien/ Brüssel

Blau, Antonia (2016): Expert interview conducted by Kristina Marion Jacobsen on 2 November 2016 in Marseille

Böhm, Andreas (2000): Theoretisches Codieren: Textanalyse in der Grounded Theory. In: Uwe Flick, Ernst von Kardorff & Ines Steinke (eds.): Qualitative Forschung. Ein Handbuch. Originalausgabe, 11. Auflage, Reinbek bei Hamburg, pp. 475–485

Bopp, Lena (1 May 2013): Rap in Marseille: Dafür ist immer noch das Sozialamt zuständig, in: Frankfurter Allgemeine Zeitung, Available online at: https://www.faz.net/aktuell/feuilleton/pop/rap-in-marsei lle-dafuer-ist-immer-noch-das-sozialamt-zustaendig-12167171. html?printPagedArticle=true#pageIndex_0 [last reviewed in June 2021]

Brüsemeister, Thomas (2000): Qualitative Forschung. Ein Überblick. Hagener Studientexte zur Soziologie, 6, Wiesbaden

Burghardt, Peter (2018): Kulturpolitik "zweckfrei fördern", in: Süddeutsche Zeitung, 14 October 2018, Available online at: https://www.sueddeutsche.de/kultur/kulturpolitik-zweckfrei-foerdern-1.4169186 [last reviewed in June 2021]

Carrijn, Flora (2019): Universities in Europe – Europe in the Universities. Reflections on the Role of University in the ECoC Context, in: Schneider, Wolfgang & Jacobsen, Kristina: Transforming Cities. Paradigms and Potential of urban development with the "European Capital of Culture", Hildesheim, pp. 25–34

Chougnet, Jean-François (2016): Expert interview conducted by Kristina Marion Jacobsen on 31 October 2016 in Marseille

Commission on Global Governance (2005): Our global neighbourhood. The report of the Commission on Global Governance, Oxford, Available online at: http://www.gdrc.org/u-gov/global-neighbourhood/chap1.htm [last reviewed in June 2021]

Čopič, Vesna & Srakar, Andrej (2012): Cultural Governance: A literature review. Available online at: http://www.eenc.info/wp-content/uploads/2012/11/VCopic-ASrakar-cultural-governance-literature-review-final.pdf [last reviewed in June 2021]

Deutscher Bundestag (ed.) (2008): Kultur in Deutschland. Schlussbericht der Enquete-Kommission des Deutschen Bundestages, Regensburg

Dirksen, Jens (2016): Expert interview conducted by Kristina Marion Jacobsen on 17 February 2016 in Essen

Fischer, Jürgen (2016): Expert interview conducted by Kristina Marion Jacobsen on 17 February 2016 in Essen

Flick, Uwe (2006): Qualitative Sozialforschung. Eine Einführung. 4. Aufl., vollst. überarb. und erw. Neuausg., Reinbek bei Hamburg

Föhl, Patrick S. & Sievers, Norbert (2013): Kulturentwicklungsplanung. Zur Renaissance eines alten Themas in der Neuen Kulturpolitik. In: Norbert Sievers (ed.): Thema: Kulturpolitik und Planung (Jahrbuch für Kulturpolitik 13), Essen, pp. 36–82

Föhl, Patrick S. & Wolfram, Gernot (2016): Transformation im Kulturbereich. Begriffe und Beispiele. In: *Kultur und Management im*

Dialog (114), pp. 32–40. Available online at: http://www.kultur-management.net/frontend/media/Magazin_Upload/km1609.pdf [last reviewed in June 2021]

Fuchs, Ulrich (2016): Expert interview conducted by Kristina Marion Jacobsen on 25 October 2016 in Marseille

Gagnon, Alain-G.; Keil, Soeren & Mueller, Sean (eds.) (2015): Understanding Federalism and Federation, London

Glaser, Barney G. (1978): Theoretical sensitivity, Mill Valley, California

Glaser, Barney G. & Strauss, Anselm L. (2009): The discovery of grounded theory. Strategies for qualitative research, New Brunswick

Graalmann, Dirk (2010): "Komm zur Ruhr". In: Süddeutsche Zeitung, 9 January 2010. Available online at: https://www.sueddeutsche.de/kultur/europaeische-kulturhauptstadt-ruhr-komm-zur-ruhr-1.53775-2 [last reviewed in June 2021]

Große Hüttmann, Martin (ed.) (2013): Das Europalexikon. Begriffe, Namen, Institutionen, Bonn

Haselbach, Dieter; Klein, Armin; Knüsel, Pius & Opitz, Stephan (2012): Der Kulturinfarkt: von allem zu viel und überall das Gleiche. Eine Polemik über Kulturpolitik, Kulturstaat, Kultursubvention, München

Heinrichs, Werner (1997): Kulturpolitik und Kulturfinanzierung. Strategien und Modelle für eine politische Neuorientierung der Kulturfinanzierung, München

Hildenbrand, Bruno (2000): Anselm Strauss. In: Flick, Uwe; Kardorff, Ernst von & Steinke, Ines (eds.): Qualitative Forschung. Ein Handbuch. Originalausgabe, 11. Auflage. Reinbek bei Hamburg, pp. 32–42

Hildesheimer Allgemeine Zeitung (21 March 2017): Was hat die Kulturhauptstadt "RUHR.2010" bewirkt? Available online at: https://www.hildesheimer-allgemeine.de/news/article/was-hat-die-kulturhaupt stadt-ruhr2010-bewirkt.html [last reviewed in June 2021]

Hoffmann, Hilmar (1981): Kultur für Alle: Perspektiven und Modelle, Frankfurt

Initiativkreis Ruhrgebiet (2007): Zukunft Ruhr 2030. Strategiepapier, Essen

Jacobsen, Kristina (2019): Raus aufs Land. Über den Einbezug ländlicher Räume innerhalb der Initiative "Kulturhauptstadt Europas", in: Drews, Albert (ed.): Ein schöner Land! Aufgaben von Kulturpolitik und Kulturarbeit im Strukturwandel ländlicher Räume, 63. Loccumer Kulturpolitisches Kolloquium, Loccumer Protokolle Bd. 13/2018, Rehburg-Loccum, pp. 181–186

Jacobsen, Kristina (2018): Ein Jahr der "Festa" auf Malta. Valletta präsentiert sich als fragwürdige "Kulturhauptstadt Europas 2018", in: Politik und Kultur, 6/2018, p. 11. Available online at: https://www. kulturrat.de/wp-content/uploads/2018/10/puk06-18.pdf [last reviewed in June 2021]

Jacobsen, Kristina (2018): Iepen Mienskip. Die Europäische Kulturhauptstadt Leeuwarden-Fryslân 2018, in: Politik und Kultur Nr. 5/2018, p. 15 Available online at: https://www.kulturrat.de/wp-cont ent/uploads/2018/08/puk05-18.pdf [last reviewed in June 2021]

Jacobsen, Kristina (2017): Ein lehrreicher Blick nach Norden. Die Europäische Kulturhauptstadt Aarhus 2017 setzt Impulse für deutsche Bewerberstädte. In: Politik und Kultur 2/2017, p. 7. Available online at: https://www.kulturrat.de/publikationen/zeitung-pk/ausgabe-nr-022017/ [last reviewed in June 2021]

Jacobsen, Kristina (2016): Kultur für ein friedliches Miteinander. Über die diesjährige Kulturhauptstadt Europas Donostia – San Sebastián. In: Kulturpolitische Mitteilungen 2/2016 (153), pp. 18–19

Jacobsen, Kristina Marion (2009): Die drei Europäischen Kulturhauptstädte in Deutschland im Kontext der Entwicklung europäischer Kulturpolitik, Saarbrücken

Kelle, Udo (1997): Empirisch begründete Theoriebildung. Zur Logik und Methodologie interpretativer Sozialforschung, Weinheim

Kemper, Jan & Vogelpohl, Anne (eds.) (2011): Lokalistische Stadtforschung, kulturalisierte Städte. Zur Kritik einer "Eigenlogik der Städte", Münster

Kießlinger, Bea & Baumann, Benedikte (2016): Nachhaltige Effekte der Kulturhauptstadt Europas RUHR.2010, in: vhw FWS 3 / Mai – Juni 2016

Kirchner, Thomas (16 May 2019): Einfach machen, in: Süddeutsche Zeitung, SZ Spezial Europa, p. 9

Klein, Armin (2007): Der exzellente Kulturbetrieb, Wiesbaden

Knoblich, Tobias J. (2018): Möglichkeiten und Grenzen kulturpolitischer Transformation am aktuellen Beispiel Thüringen, in: Zeitschrift für Kulturmanagement, 4/1, pp. 87–102

Knoblich, Tobias J. & Scheytt, Oliver (2009): Zur Begründung von Cultural Governance, in: Aus Politik und Zeitgeschichte 2009 (08/2009), pp. 34–40. Available online at: http://www.bpb.de/apuz/32173/politische-steuerung [last reviewed in June 2021]

Kolb, Andreas & Scheytt, Oliver (2010): Eine Kulturhauptstadt von allen und für alle. Oliver Scheytt im Gespräch, in: Neue Musikzeitung 4/2010 – 59. Jahrgang, Available online at: https://www.nmz.de/artikel/eine-kulturhauptstadt-von-allen-und-fuer-alle [last reviewed in June 2021]

Kohler-Koch, Beate & Eising, Rainer (eds.) (2009): The transformation of governance in the European Union (transfer to digital print 2007), London

Kollmorgen, Raj (1996): Schöne Aussichten? Eine Kritik integrativer Transformationstheorien. In: Kollmorgen, Raj; Reißig, Rolf & Weiß, Johannes (eds.): Sozialer Wandel und Akteure in Ostdeutschland. Empirische Befunde und theoretische Ansätze; [Ergebnisse der sozialwissenschaftlichen Transformationskonferenz "Sozialer Wandel und Akteure in (Ost-)Deutschland" ... am 12. und 13. Dezember 1994 in Berlin]. Opladen, pp. 281–332

Könneke, Achim (2011): Kulturpolitik in Großstädten. Kontext und Konzept des kulturkonzept.freiburg, in: Hessische Vereinigung für Volkskunde & Markus Morr (eds.): Kultur und Politik. Aspekte kulturwissenschaftlicher und kulturpolitischer Spannungsfelder, 47 (2011). (Hessische Blätter für Volks- und Kulturforschung, N.F. 47), Marburg, pp. 114–128

Kuckartz, Udo; Ebert, Thomas; Rädiker, Stefan & Stefer, Claus (2009): Evaluation online. Internetgestützte Befragung in der Praxis, Wiesbaden

Leydolt-Fuchs, Pia (2016): Expert interview conducted by Kristina Marion Jacobsen on 3 November 2016 via Skype

Liehr, Günter (2013): Marseille. Portrait einer widerspenstigen Stadt, Zürich

Löw, Martina (2009): Erwartungen an nationale Stadtentwicklungspolitik?, in: Bundesministerium für Verkehr, Bau & Stadtentwicklung (eds.): Nationale Stadtentwicklungspolitik, Berlin, pp. 54–55. Available online at: https://www.nationale-stadtentwicklungs politik.de/NSP/SharedDocs/Publikationen/DE_NSP/positionen.pdf; jsessionid=68AAE1D5F680B3FE1F28D78C6E2DBF54.live11 293?_blob=publicationFile&v=1 [last reviewed in June 2021]

Löw, Martina (2008): Eigenlogische Strukturen – Differenzen zwischen Städten als konzeptuelle Herausforderung, in: Berking, Helmuth & Löw, Martina (eds.): Die Eigenlogik der Städte. Neue Wege für die Stadtforschung, Darmstadt, pp. 33–53

Lueger, Manfred (2007): Grounded Theory, in: Buber, Holzmüller (ed.) – Qualitative Marktforschung, pp. 189–205

Magel, Eva-Maria (2009): Projekt "Eigenlogik der Städte". Neue Blicke auf die Stadt, in: Frankfurter Allgemeine Zeitung, 2 June 2009 Available online at: https://www.faz.net/aktuell/rhein-main/region-und-hessen/projekt-eigenlogik-der-staedte-neue-blicke-auf-die-stadt-1810 679.html [last reviewed in June 2021]

Maisetti, Nicolas (2013): Marseille 2013 Off: l'institutionnalisation d'une critique?, in: Faire Savoirs No. 10, décembre 2013, pp. 59–68

Mandel, Birgit (2017): Die Trennlinien zwischen E- und U-Kultur auflösen. Kulturwirtschaft in Deutschland. Fakten, Diskurse und Perspektiven des kulturunternehmerischen Schaffens, in: Hausmann, Andrea & Heinze, Anne: Cultural Entrepreneurship – Gründen in der Kultur- und Kreativwirtschaft, Wiesbaden, pp. 13–26

Mandel, Birgit & Renz, Thomas (2010): Barrieren der Nutzung kultureller Einrichtungen. Eine qualitative Annäherung an Nicht-Besucher, Hildesheim

Megerle, Heidi (2017): Centre-ville pour tous oder Kulturhauptstadt für Ausgewählte? Aktuelle Stadterneuerungsprozesse in Marseille und ihre Konsequenzen für prekarisierte Bevölkerungsgruppen, in: Altrock, Uwe & Kunze, Ronald (eds.): Stadterneuerung und Armut. Jahrbuch Stadterneuerung 2016, Wiesbaden, pp. 343–372

Megerle, Heidi (2011): Lebenszyklen und Transformationsprozesse eines städtischen Boulevards – die Rue de la République in Marseille, in: Proceedings REAL CORP 2011 Tagungsband 18–20 May 2011, Essen, pp. 533–542

Megerle, Heidi (2008): Von der "ville en crise" zur Metropolitan European Growth Area: aktuelle Transformationsprozesse der Metropolregion Marseille-Aix-en-Provence, in RuR I/2008, pp. 23–35

Meinefeld, Werner (2000): Hypothesen und Vorwissen in der qualitativen Sozialforschung, in: Uwe Flick, Ernst von Kardorff & Ines Steinke (eds.): Qualitative Forschung. Ein Handbuch. Originalausgabe, 11. Auflage, Reinbek bei Hamburg, pp. 265–275

Meinefeld, Werner (1995): Realität und Konstruktion. Realität und Konstruktion – Erkenntnistheoretische Grundlagen einer Methodologie der empirischen Sozialforschung, Wiesbaden

Mittag, Jürgen (2019): The legacy of "RUHR.2010": Memories of the European Capital of Culture 2010, in: Schneider, Wolfgang & Jacobsen, Kristina: Transforming Cities. Paradigms and Potential of urban development with the "European Capital of Culture", Hildesheim, pp. 197–208

Mittag, Jürgen (2016): Expert interview conducted by Kristina Marion Jacobsen on 16.03.2016 via telephone

Mittag, Jürgen (2013): The changing concept of the European Capitals of Culture between the endorsement of European identity and city advertising, in: Patel, Kiran Klaus: The Cultural Politics of Europe. European Capitals of Culture and European Union since the 1980s, London, pp. 39–54

Mittag, Jürgen (2012): Kulturhauptstadt Europas. Eine Idee – viele Ziele – begrenzter Dialog, in: Ernst, Thomas; Heimböckel, Dieter: Verortungen der Interkulturalität. Die Europäischen Kulturhauptstädte Luxemburg und die Großregion (2007), das Ruhrgebiet (2010) und Instanbul (2010), Bielefeld

Mittag, Jürgen (2008): Die Idee der Kulturhauptstadt Europas. Vom Instrument europäischer Identitätsstiftung zum tourismusträchtigen Publikumsmagneten, in: Mittag, Jürgen (ed.): Die Idee der Kulturhauptstadt Europas. Anfänge, Ausgestaltung und Auswirkungen Europäischer Kulturpolitik, Essen, pp. 55–96

Neundlinger, Barbara (2019): EU-Kulturpolitik im Zeichen von Dialog, Nachhaltigkeit und Zusammenhalt. Die europäische Agenda für Kultur und Arbeitsplan für Kultur 2019–2022, in: Kulturpolitische Mitteilungen 164, I/2019, pp. 30–31

Nohlen, Dieter (ed.) (1998): Politische Begriffe, Lexikon der Politik 7, München

Palmer, Robert & Rae Associates (2004): European Cities and Capitals of Culture, Parts I and II, Brüssel

Patel, Kiran Klaus (2013): Introduction, in: Patel, Kiran Klaus (ed.): The Cultural Politics of Europe. European Capitals of Culture and European Union since the 1980s, Abingdon, pp. 1–15

Petzinger, Tana; Schulte, Stephan; Scheytt, Oliver & Tum, Carsten (2009): Regional Governance in der Metropole Ruhr, in: Ludwig, Jürgen; Mandel, Klaus; Schwieger, Christopher & Terizakis, Georgios (eds.): Metropolregionen in Deutschland, Baden-Baden, pp. 144–157

Pleitgen, Fritz (2010): Ruhr. Vom Mythos zur Marke, in: Frohne, Julia; Langsch, Katharina; Pleitgen, Fritz & Scheytt, Oliver (eds.): Ruhr. Vom Mythos zur Marke. Marketing und PR für die Kulturhauptstadt Europas RUHR.2010, Essen, pp. 6–8

Prante, Martina (5 December 2015): Schön wär's ja doch. Will, soll, kann Hildesheim 2025 Kulturhauptstadt werden? In: Hildesheimer Zeitung, p. 11. Available online at: https://www.uni-hildesheim.de/uploads/media/2015_12_05_hiaz-kupo-kulturhauptstadt-schneider-marseille-dammann-teilhabe-kultur.pdf [last reviewed in June 2021]

Reichle, Leon (15 January 2019): Die Evakuierten, in: ak analyse & kritik, Nr. 645, Available online at: https://www.akweb.de/ak_s/ak645/11.htm [last reviewed in June 2021]

Renner, Tim (2016): Wie wichtig wir sind, bestimmen wir selbst. Notizen zur aktuellen Kulturpolitik, in: Norbert Sievers, Tobias J. Knoblich & Patrick S. Föhl (eds.): Thema: Transformatorische Kulturpolitik (Jahrbuch für Kulturpolitik, Band 15), Bielefeld, pp. 369–373

Renz, Thomas (2015): Nicht-Besucherforschung: Die Förderung kultureller Teilhabe durch Audience Development, Bielefeld

Renz, Thomas (2014): Besuchsverhindernde Barrieren im Kulturbetrieb, in: Mandel, Birgit & Renz, Thomas: Mind the gap! Zugangsbarrieren zu kulturellen Angeboten und Konzeptionen niedrigschwelliger Kulturvermittlung, Hildesheim, pp. 22–34

Ruhr.2010 GmbH i.L. (2012): "Essen für das Ruhrgebiet": Ruhr.2010 Kulturhauptstadt Europas. Programm und Wirkung, Essen

Rydzy, Edda & Griefahn, Monika (2014): Natürlich wachsen. Erkundungen über Mensch, Natur und Wachstum aus kulturpolitischem Anlass, Wiesbaden

Schäfer, Sandra (2017): Kulturhauptstadt ist kein Wanderzirkus (7 May 2017). Available online at: https://kulturfuechsin.com/at/interview-kulturhauptstaedte-2017-cap-cult/ [last reviewed in June 2021]

Scheytt, Oliver (2018): Expert interview conducted by Kristina Marion Jacobsen on 12 March 2018 via e-mail

Scheytt, Oliver (2006): Kulturhauptstadt-Bewerbung als Regional Governance, in: Kleinfeld, Ralf; Plamper, Harald; Huber, Andreas (eds.): Regional Governance. Steuerung, Koordination und Kommunikation in regionalen Netzwerken als neue Formen des Regierens, Band 1 von 2, Göttingen, pp. 207–216

Schneider, Gerald & Aspinwall, Mark (eds.) (2001): The rules of integration. Institutionalist approaches to the study of Europe, Manchester Univ. Press (European Policy Research Unit series), Manchester

Schneider, Wolfgang (2019): Avenir ou impasse esthétique? Transformation als Gegenstand von Forschung und Lehre in den Kulturwissenschaften, in: Schneider, Wolfgang; Butel, Yannick; Bärwolff, Theresa; Suzanne, Gilles (eds.): Dispositive der Transformation. Kulturelle Praktiken und künstlerische Prozesse, Hildesheim, pp. 176–179

Schneider, Wolfgang (2016): "Seismographs", "Watch Dogs" or "Change Agents"? Artistic invention and cultural policy in progress of social transformation, in: Vickery, J.P. (ed.): The law, social justice and global development journal, No. 1/2016, pp. 1–8. Available online at: https://warwick.ac.uk/fac/soc/law/elj/lgd/2016-1/schneider_finalfinall.pdf [last reviewed in June 2021]

Schneider, Wolfgang & Gad, Daniel (eds.) (2014): Good Governance for Cultural Policy: An African-European Research about Arts and Development, Frankfurt

Schneider, Wolfgang (2011): Kultur plus Politik gleich Kulturpolitik? Lebenskunst als gesellschaftlicher Auftrag, in: Hessische Vereinigung für Volkskunde & Markus Morr (eds.): Kultur und Politik. Aspekte kulturwissenschaftlicher und kulturpolitischer

Spannungsfelder. (Hessische Blätter für Volks- und Kulturforschung, N.F. 47), Marburg, pp. 11–23

Schneider, Wolfgang (2008): Vom Export zum Netzwerk, vom Event zur Intervention. Zum Wandel Auswärtiger Kulturpolitik, in: Schneider, Wolfgang: Auswärtige Kulturpolitik. Dialog als Auftrag – Partnerschaft als Prinzip, Essen, pp. 13–31

Schultheis, Karl (2016): Expert interview conducted by Kristina Marion Jacobsen on 17 February 2016 in Düsseldorf

Schuppert, Gunnar Folke & Zürn, Michael (2008): Governance in einer sich wandelnden Welt, Politische Vierteljahresschrift Sonderheft, 41/2008, Wiesbaden

Schwencke, Olaf (2019): Europe's Urban Culture. The Basic Value for Shaping Democracy, in: Schneider, Wolfgang & Jacobsen, Kristina: Transforming Cities. Paradigms and Potential of Urban Development with the "European Capital of Culture", Hildesheim, pp. 53–58

Schwencke, Olaf (2015): Europa. Kultur.Politik. Die kulturelle Dimension im Unionsprozess, Bonn

Schwencke, Olaf (2010): Das Europa der Kulturen – Kulturpolitik in Europa (dritte Auflage), Essen

Schwencke, Olaf (2005): Kulturhauptstädte Europas. Die Zukunft der Stadt als europäische Lebensform, in: Kulturpolitische Mitteilungen Nr. 111, IV 2005, pp. 36–38

Schwencke, Olaf & Rydzy, Edda (2004): Kulturstädte als Hefe europäischer Entwicklung und Integration. 18 deutsche Städte bewerben sich um den Titel Kulturhauptstadt Europas 2010, in: Kulturpolitische Mitteilungen Nr. 104, I/2004, pp. 8–9

Schwencke, Olaf (1996): Kulturpolitik im Spektrum der Gesellschaftspolitik, in: Aus Politik und Zeitgeschichte 41, pp. 3–11

Seawright, Jason & Gerring, John (2008): Case Study Selection Techniques in Case Study Research. A Menu of Qualitative and Quantitative Options. In: Political Research Quarterly 61 (2), pp. 294–308. Available online at: http://journals.sagepub.com/doi/pdf/10.1177/1065912907313077 [last reviewed in June 2021]

Siebel, Walter (1987): Vorwort zur deutschen Ausgabe, in: Saunders, Peter (ed.): Soziologie der Stadt, Frankfurt a.M., pp. 9–13

Sievers, Norbert; Föhl, Patrick S. & Knoblich, Tobias
J. (eds.) (2016): Jahrbuch für Kulturpolitik 2015/2016, Bd. 15,
Thema: Transformatorische Kulturpolitik, Bielefeld

Simmel, Georg (2006): Die Großstädte und das Geistesleben, Berlin

Singer, Otto (2010): EU-Kulturpolitik nach Lissabon. In: Infobrief des
Wissenschaftlichen Dienstes des Deutschen Bundestages (30 March
2010) Available online at: https://www.bundestag.de/blob/191460/
5adf6e6118e415fcfc268ccb3796e265/eu-kulturpolitik_nach_lissabon-
data.pdf [last reviewed in June 2021]

Steinke, Ines (2000): Gütekriterien qualitativer Forschung, in: Flick, Uwe;
Kardorff, Ernst von & Steinke, Ines (eds.): Qualitative Forschung. Ein
Handbuch. Reinbek bei Hamburg, pp. 319–331

Stratenschulte, Eckart D. (2018): Mehr als Geld, in: Cicero (23 April
2018), Available online at: https://www.cicero.de/aussenpolitik/europa-
eu-identitaet-merkel-macron-werte [last reviewed in June 2021]

Strauss, Anselm & Corbin, Juliet (2013): Grounded Theory
Methodology. An Overview, in: Norman K. Denzin & Yvonna
S. Lincoln (eds.): Strategies of Qualitative Inquiry. Los Angeles,
pp. 158–183

Strauss, Anselm & Corbin, Juliet (2010): Grounded theory. Grundlagen
qualitativer Sozialforschung. Weinheim

Suzanne, Gilles (2016): Expert interview conducted by Kristina Marion
Jacobsen on 1 November 2016 in Marseille

Taylor, Ian; Evans, Karen & Fraser, Penny (1996): Tale of two
cities: Global change, local feeling and everyday life in the North of
England. A study in Manchester and Sheffield, London

United Nations (1987): Report of the World Commission on
Environment and Development: Our Common Future. Available
online at: https://sustainabledevelopment.un.org [last reviewed in
June 2021]

Weber, Max (2009): Die Stadt (1921), in: Weber, Max: Gesamtausgabe,
Studienausgabe: Band I/22,1: Wirtschaft und Gesellschaft, Tübingen

Weizsäcker, Richard von (1987): Die politische Kraft der Kultur, Reinbek

Wessels, Wolfgang (2008): Das politische System der Europäischen
Union, Wiesbaden

Woker, Martin (2 July 2015): Marseilles Aufstieg. Die Grenzen der Stadtentwicklung. In: Neue Zürcher Zeitung vom 2 July 2015. Available online at: http://www.nzz.ch/international/europa/meersicht-und-die-macht-der-maerkte-1.18572780, [last reviewed in June 2021]

Zimmermann, Olaf (2018): Die Perspektive wechseln. Zur kulturellen Dimension der Nachhaltigkeitsdebatte, in: Politik und Kultur 1/2018, p. 17

Legal documents and other EU documents

All links listed here were last accessed in June 2021. They are listed according to the order in which they are mentioned in this work.

– Link 1 Opinion of the Committee on Culture and Education for the Committee on Budgets on the Interim report on MFF 2021–2027 – Parliament's position in view of an agreement (COM(2018)0322 – 2018/0166R(APP)) Available online at: http://www.europarl.europa.eu/doceo/document/CULT–AD–627000_EN.pdf

– Link 2 Resolution of the Ministers responsible for Cultural Affairs, meeting within the Council, of 13 June 1985 concerning the annual event "European City of Culture" (85/C 153/02). Available online at: https://eur-lex.europa.eu/legal-content/EN/TXT/PDF/?uri=CELEX:41985X0622&rid=2

– Link 3 Decision 1419/1999/EC of the European Parliament and of the Council of 25 May 1999 establishing a Community Action for the European Capital of Culture event for the years 2005 to 2019. Available online at: http://eur-lex.europa.eu/legal-content/EN/TXT/?uri=CELEX:31999D1419

– Link 4 Address given by Jacques Delors to the European Parliament (17 January 1989), in: Bulletin der Europäischen Gemeinschaften. 1989, n° Sonderbeilage 6/1989, Luxemburg, Amt für amtliche Veröffentlichungen der Europäischen Gemeinschaften, Available online at: https://www.cvce.eu/content/publication/2003/8/22/b9c06b95-db97-4774-a700-e8aea5172233/publishable_en.pdf

– Link 5 Decision No 445/2014/EU of the European Parliament and of the Council of 16 April 2014 establishing a Union action for the European Capitals of Culture for the years 2020 to 2033 and repealing

Decision No 1622/2006/EC Available online at: https://eur-lex.europa.eu/
legal-content/EN/TXT/PDF/?uri=CELEX:32014D0445&from=EN

– Link 6 Decision 1622/2006/EC of the European Parliament and of
the Council of 24 October 2006 establishing a Community action
for the European Capital of Culture event for the years 2007 to 2019
Available online at: http://eur-lex.europa.eu/legal-content/EN/TXT/
HTML/?uri=CELEX:32006D1622&from=DE

– Link 7 Barroso, José Manuel: Foreword (2009), in: European
communities (eds.): European Capitals of Culture: The Road to
Success from 1985 to 2010, Luxemburg, S. 1 Available online
at: https://ec.europa.eu/programmes/creative-europe/sites/creative-
europe/files/library/capitals-culture-25-years_en.pdf

– Link 8 REGIONS 2020. An Assessment of Future Challenges for EU
Regions, Commission Staff Working Document, November 2008.
Available online at: http://ec.europa.eu/regional_policy/sources/
docoffic/working/regions2020/pdf/regions2020_en.pdf

– Link 9 Treaty of Lisbon amending the Treaty on European Union and
the Treaty establishing the European Community, signed at Lisbon,
(2007/C 306/01), 13 December 2007. Available online at: https://eur-lex.
europa.eu/legal-content/EN/TXT/?uri=CELEX:C2007/306/01

– Link 10 Consolidated versions of the Treaty on European Union and
the Treaty on the Functioning of the European Union (2010/C 83/01).
Available online at: https://eur-lex.europa.eu/legal-content/EN/TXT/
?uri=CELEX:C2010/083/01

– Link 11 Euro-Mediterranean Association Agreements. Available
online at: https://eur-lex.europa.eu/legal-content/EN/TXT/?uri=LE
GISSUM%3Ar14104

– Link 12 Ex-post Evaluation of the 2013 European Capitals
of Culture, Final Report for the European Commission DG
Education and Culture, December 2014. Available online
at: https://op.europa.eu/de/publication-detail/-/publication/
92ec5895-b37b-11e5-8d3c-01aa75ed71a1

– Link 13 Evaluation reports and other European Commission
documents on the European Capital of Culture Overview. available
online at: https://ec.europa.eu/culture/policies/culture-cities-and-
regions/european-capitals-culture

– Link 14 Communication from the Commission to the European
  Parliament, the European Council, the Council, the European
  Economic and Social Committee and the Committee of the
  Regions: A New European Agenda for Culture, COM(2018) 267
  final (22.05.2018). Available online at: https://eur-lex.europa.eu/legal-
  content/EN/TXT/?uri=COM%3A2018%3A267%3AFIN

– Link 15 European Commission: European governance – A white paper,
  COM/2001/0428 final (12.10.2001). Available online at: http://eur-lex.
  europa.eu/legal-content/EN/TXT/?uri=CELEX:52001DC0428

– Link 16 Council conclusions on the Work Plan for Culture
  2019–2022 (21.12.2018), 2018/C 460/10. Available online
  at: https://eur-lex.europa.eu/legal-content/EN/TXT/?uri=
  CELEX%3A52018XG1221%2801%29

– Link 17 European Parliament, Directorate-General for Internal
  Policies, Policy Department B Structural and Cohesion Policies, note
  (2012): "The culture strand of the creative Europe programme
  2014 – 2020". Available online at: http://www.europarl.europa.eu/
  RegData/etudes/note/join/2012/495825/IPOL-CULT_
  NT%282012%29495825_EN.pdf

Other links

– Link 18 http://www.proloewe.de/eigenlogik

– Link 19 https://www.bundesregierung.de/breg-de/service/bulletin/
  regierungserklaerung-von-bundeskanzlerin-dr-angela-merkel-795942

– Link 20 https://www.parliament.uk/about/living-heritage/building/
  palace/architecture/palacestructure/churchill/

– Link 21 https://wissenschaft.hessen.de/sites/default/files/media/hmwk/
  loewe_abschlussbericht_eigenlogik.pdf

– Link 22 https://kulturfuechsin.com/at/
  interview-kulturhauptstaedte-2017-cap-cult/

– Link 23 https://www.sueddeutsche.de/kultur/
  europaeische-kulturhauptstadt-ruhr-komm-zur-ruhr-1.53775-2

– Link 24 https://www.friesland.nl/de/kulturhauptstad-2018/uber/haufig-
  gestellte-fragen/hintergrund

– Link 25 https://www.rvr.ruhr/fileadmin/user_upload/01_RVR_Home/
02_Themen/Kultur/Evaluationsbericht_Ruhr.2010.pdf

– Link 26 https://www.e-c-c-e.de/european-centre-for-creative-economy.html

– Link 27 http://www.bundestag.de/blob/189558/21543d1184c1f627412
a3426e86a97cd/charta-data.pdf

– Link 28 https://www.centrevillepourtous.asso.fr/

– Link 29 Declaration on Cultural Policies, World Conference on
Cultural Policies, Mexico City, 26 July – 6 August 1982. Available
online at http://www.unesco.org/new/en/culture/themes/dynamic-
content-single-view/news/mexico_city_declaration_on_cultural_
policies_world_conferen/

– Link 30 UN resolution "Transforming our world: the 2030 Agenda
for Sustainable Development", [adopted on 25 September 2015 (A/
70/L.1)] 70/1. Available online at https://www.un.org/en/development/
desa/population/migration/generalassembly/docs/globalcompact/A_
RES_70_1_E.pdf

– Link 31 http://www.lamarseillaise.fr/alpes/societe/
41092-il-ne-suffit-pas-de-bien-faire-son-travail

– Link 32 http://www.biennale-cirque.com/fr/

– Link 33 http://www.mp2018.com/

– Link 34 http://ec.europa.eu/environment/europeangreencapital/index_
en.htm

– Link 35 https://www.essen.de/rathaus/europa/eu_projekte/european_
green_capital.de.html

– Link 36 https://www.youtube.com/watch?v=ta2jot9JhU0

– Link 37 https://www.youtube.com/watch?v=CEg1jMeTIjQ

Afterword

The publication of an article by investigative journalist Uwe Ritzer in the *Süddeutsche Zeitung* on 3 December 2020 suddenly cast the ECoC initiative in a different light. Ritzer also focused on sustainable structures – but in a completely different way. He revealed continuing connections between the selection panel, a network of consultants, and the ECoC actors; these connections cast doubt on the credibility of the title award. Essentially, Ritzer drew attention to two issues: (1) the possible bias of at least one juror in the selection of the city of Chemnitz as ECoC 2025; and (2) the lack of transparency of the initiative, which allows some consultants to conclude contracts for high fees below the radar of public awareness. The journalist's research appears to be trustworthy and corresponds to the facts. It reveals that there are design flaws in the ECoC initiative that have resulted in "sustainable" grievances.

The competition in Germany for the ECoC title for 2025 not only produced countless "identity discourses", "cultural policy potentials", "municipal-regional reflections" and "new governance approaches", but also revealed the dark side of these noble goals. Just as RUHR.2010 became the new benchmark in ECoC history in terms of its budget, its programme design and, above all, its regional scope, the German candidate cities for 2025 also wanted to set new standards. It had long since become standard practice for cities to include their regions in their programmes and also in the years of preparation and follow-up. No city would have proceeded without this regional aspect, since a larger area makes it possible to involve even more people in the project and to promote (or find) even more cultural treasures. So, naturally, little Zittau made its bid together with the regions in its two neighbouring countries. Nuremberg included the entire metropolitan region, which is larger than the federal state of Hesse. And, in Hildesheim, the whole bid was made under the title of "a European province of culture" anyway. The "capital of culture" had thus become a cultural region. Since the topic of "culture in rural areas" has long been in vogue in German cultural policy, this further development was generally welcomed. Who would have anything negative to say about the pretty rural cultural projects that gave the cities' bids a touch of country charm, closeness to nature, originality and authenticity, in addition to their urban profile?

Over time, the inclusion of the surrounding region was added to the EU's list of criteria, as were further requirements in the area of urban development. The quantitative scope of the legal developments alone shows how much the demands have grown over the years for a city to assert itself as an ECoC against competing cities: While the first resolution of 1985 comprised only one page, the resolution has become increasingly comprehensive over time and now comprises 12 pages. The fact that the demands on the cities participating in the competition increased was generally welcomed: even more goals had become possible. Thus, over time, the ECoC initiative became a multifunctional mega-project to which the most diverse stakeholders could subscribe.

No one pointed out the price that this development brought with it: not only more money, but also more know-how became necessary. But among the euphorically proclaimed slogans of the candidate cities ("Here Now Everyone for Europe",[47] or "How Candidate Cities Can Contribute the Redesign of Europe"[48]) there was no room for critical voices.

Because each candidate city would be given a budget of tens of millions of Euros if they were to win the title, consultants rallied around them and offered their services, as they always do when large sums of money are involved. And since an ECoC bid is a complex and specialised matter, the expertise for which can be acquired only through personal experience, the consultants were also needed. They form a manageable group of about 30 persons who move from ECoC to ECoC and make their knowledge available. Some cynically call their business a "consulting rodeo",[49] while others consider it crucial insider knowledge that can help a city to win. It is indisputable that there was a need for advice in the cities about very extensive requirements of the bidding process. Certainly, consultants can

---

47 Motto of the candidate city Hanover: see link (a).
48 Title of a panel at the ECoC Conference "under construction" in Magdeburg, 22–24 March 2018: see link (b).
49 "A consulting rodeo with all its negative implications, such as recommending each other to others, and passing posts and contracts" (Kauffmann 2020).

be helpful when it comes to the "view from the outside" in a demanding, year-long process that is managed by a small bid team that often has little experience of large-scale projects in a European context. But the question is whether the scale and the amount of the fees are justified. Is it justifiable – not least from an environmental point of view – to have consultants flown in from all over Europe? Was there really no one in Saxony or Germany with the appropriate expertise for Chemnitz? There is no question that some advisors are committed, stay on site, have a close connection with the cities and assist them in the long term. But there are also those who operate exclusively in the background and move from location to location, presenting their supposedly unique concepts and pressuring the cities to pay their high fees.

The shadow economy of the consultants is made possible by the lack of transparency in the processes followed in awarding the ECoC title. As the *Süddeutsche Zeitung* revealed in several articles in December 2020 and January 2021, several consultants in different candidate cities received six-figure sums, working for competing candidate cities at the same time and passing consulting contracts to each other (Ritzer 2020). There were also dubious connections with the selection panel. The consultants were never based in one of the candidate cities and only in exceptional cases did they become publicly involved in the discourse. In any event, the advancement of the initiative was not a priority for them, even if this was sometimes presented differently.

Ritzer explains the entanglements of the selection jury as follows:

> These conditions cannot be justified by the fact that there is only a small circle of experts who know their way around this topic. If this is indeed the case, two questions arise: Firstly, whether the professional requirements for the ECoC label are not fundamentally wrong and completely excessive, so that candidate cities are forced to buy in expensive people from outside. The second question is whether the circle of these people is deliberately kept so manageable precisely because they can then more conveniently offer each other the contracts (Ritzer 2021: 4).

The financial aspect of this "consultant-gate" alone is a scandal: While competitions were held in the bidding process in which artists and cultural workers could win the symbolic sum of 2025 Euros (appropriate for the ECoC year 2025), consultants received multiples of this sum. Cities could decorate themselves outwardly using the micro-financing provided for colourful, sympathetic cultural projects, and the cultural scene was supposed to be happy about the small gifts received from the bidding team. What was not communicated to the outside world, however, was that, behind closed doors, consultants sometimes received fees that were more than 50 times as high.[50]

In addition to the disclosure of the consultancy deals, the conflicts of interest of a member of the ECoC selection jury were also revealed. One juror is the CEO of a cultural centre listed as a cooperation project in Chemnitz' bid book.[51] Thus, by voting for Chemnitz, the juror directly benefited from his own decision.

The political handling of a suspected conflict of interest

This alleged bias became known in December 2020, i.e. only a few weeks after the European selection panel's designation of Chemnitz on 28 October 2020. At that time, however, the formal designation by the Federal Republic of Germany, represented by the Conference of Ministers of Education and Cultural Affairs (Kultusministerkonferenz or "KMK"), was still pending. A troubling situation now ensued: several politicians raised their concerns, first and foremost the Bavarian Minister of Education and Cultural Affairs, Bernd Sibler, who was chairing the Conference of Ministers of Culture at the time. But he was immediately silenced by various media: the Bavarians were labelled "sore losers" because they had thought they would win with Nuremberg. But who else but the defeated candidate cities could have made a qualified statement on this matter? After nine months of Covid, the cultural sector had enough to do simply trying to survive, so it could not address the complex matter of the ECoC bidding process. Consequently, only those who were themselves involved in the matter, i.e. the candidate cities, were well versed in it.

Sibler's complaint led to the KMK's decision being postponed so that they could investigate the accusations of bias. This was already seen as a scandal

---

50 See Ritzer (2020).
51 See link (c).

because never before had an ECoC not been confirmed by the competent national bodies after being nominated by the selection panel.

At the beginning of 2021, the chairmanship of the KMK moved from Bernd Sibler to Berlin's Senator for Culture, Klaus Lederer. Under his chairmanship, the KMK confirmed the appointment of Chemnitz on 11 January 2021. Another option was probably not conceivable, because the selection procedure does not provide for a replacement. Thus, another candidate city could not have easily taken Chemnitz' place, and, furthermore, such a move might have led to the politically unwelcome East–West debate. The KMK succinctly stated that, at the meeting, "important questions raised in the media in connection with the selection of the European Capital of Culture 2025 [were] clarified".[52] However, what was clarified, and how, remains a secret to this day.

The chairperson of the jury was not available to answer questions from the media and academia. The European Commission (EC) only gave vague rather than concrete answers to the reasonable questions about how the accusations of bias against the juror in question could be cleared up. Enquiries to the KMK and individual culture ministers repeatedly referred to the KMK's enigmatic press release and explained that the meeting was "confidential". Accordingly, no minutes of the meeting were released.

This raises pressing questions about both the proportionality and the appropriateness of the failure to communicate and the citizens' right to access information under the Freedom of Information Act. After all, this was not an explosive intelligence operation, but a cultural competition. Questions about the outcome of a multi-million project using taxpayers' money cannot be classified as confidential in this case, but must be addressed in a completely transparent manner. In this respect, the reactions of some leading journalists were justified: the political procedure was labelled a "tour de force in terms of backroom mentality" (Hansen 2020), where "morals [...] seem to have been left a bit behind on the way of Chemnitz as European Capital of Culture 2025".[53]

---

52 See link (d).
53 See link (e).

There were also talks between the EC and the chairperson of the European Parliament's Culture Committee on the allegations in question. Here, too, those responsible could only unanimously claim that the allegations had been adequately addressed. Not even the European Parliament, whose duty it is to control the executive, has commented on the lack of transparency in the allocation of millions of Euros of taxpayers' money.

The handling of the accusation of bias against a member of the selection panel is a political scandal that has caused lasting damage to the ECoC initiative. It is unclear whether the selection panel simply overlooked the involvement of a juror in part of the candidate city's bid book or deliberately kept quiet about it. Neither possibility casts a good light on the renowned cultural experts who are entrusted with awarding titles with far-reaching consequences for cities and regions. After the publication of the accusations, there should at least have been a public reaction from the EU side, i.e. from the selection panel or the responsible office in the EC. This would have been a good opportunity to counter the accusation of lack of transparency in the selection process (even though overdue). Their silence, however, reinforces the suspicion that there were conflicts of interest in the awarding of the ECoC title – and perhaps not only in Chemnitz.

The costly activities of a few advisors in the background and the silence about critical aspects of the award process are the climax of a continuing lack of transparency, and have led to the fall of idealistic, noble goals. One of the design flaws of the initiative is that there are no waiting periods when stakeholders wish to take on a different role, for example, when former jurors become consultants, as is the case in the business world when similarly high sums are involved. So far, smooth transitions between activities as a juror, advisor and cultural manager or even working in these positions at the same time have been a kind of normal state of affairs, so that no one has objected.

Open questions

Basically, the question is how to ensure transparency in the ECoC selection process. What exactly are the selection criteria for the jurors? Who verifies the independence and integrity of the jurors? What happens in the case of irregularities? The current regulations only specify the composition of the selection panel[54] – but not what the selection of the jurors is based on and who assesses the panel and, if necessary, intervenes if irregularities occur.

It is also incomprehensible why the city visits that the German candidate cities presented for the final selection in October 2020, which took place completely online due to the Covid-19 pandemic, are being kept under wraps. After all, they were financed with hundreds of thousands of Euros that come almost exclusively from the public purse. The digital city visits do not contain state secrets, but merely summarise the bids of the respective cities in video format. What is so strictly confidential about a video in which a city presents its cultural capital and potential?

It must also be asked why the names of the jury members are announced at short notice before the title is awarded and why there is no provision in principle for the jurors to be available for interviews. Last, but not least, the public has a right to know exactly how the allegations of conflicts of interest in the awarding of the title to Chemnitz were addressed. And finally: What concrete consequences have resulted or will result from the allegations of bias and lack of transparency against the ECoC initiative that have been discussed since December 2020?

When asked, the EC stated that the juror in question had left the selection panel "for professional reasons". There was no press release about this change. The names of the jury members do not appear on the EC website;[55] on request, one learns that they can be found in a selection report. Calls for more transparency have so far been ignored. In February 2021, the German KMK approached the responsible EU Commissioner for Culture, Marija Gabriel, and called for a review of the EU competition framework, including transparency and compliance regulations. Now the topic will be

---

54 The independent European selection panel is formally composed as follows: The European Commission, the European Council and the European Parliament each appoint three members for three years. The European Committee of the Regions, the Standing Conference of the Ministers of Education and Cultural Affairs of the Länder in the Federal Republic of Germany and the Federal Government (the Federal Government Commissioner for Culture and the Media and the Federal Foreign Office) each appoint one jury member. See Decision No 445/2014/EU, Art. 6: link (f).

55 See link (g).

on the agenda of the European Parliament's Culture Committee. It remains to be seen what concrete improvements will now be made.

Misguided developments of the original idea

Only a few years ago, there were calls for the EU's bid book criteria to demand more from cities (see Jacobsen 2009: 47). But now a kind of cobra effect[56] has occurred. The catalogue of criteria has become so demanding that smaller cities no longer have the opportunity to participate in the competition on their own – unless they allow themselves to be coached by external consultants. Even Chemnitz, which is not really small and does not lack experience with international cooperation projects, spent about 650,000 Euros on external consultants, according to information it made available. But does this still correspond to the original idea of the ECoC, if a small group of ECoC consultants moves from city to city, offering their "off-the-shelf originality"?[57]

---

56 "A perverse incentive is an incentive that has an unintended and undesirable result that is contrary to the intentions of its designers. The cobra effect is the most direct kind of perverse incentive, typically because the incentive unintentionally rewards people for making the issue worse" See link (h).

57 See the illustration in the bid book of the candidate city Hanover: "And then a consultant from an agency comes to see you and tells you that the deadline for bids for the title of European Capital of Culture 2025 is fast approaching. A very, very important deadline, he says. But you needn't worry. He's already written lots of Bid Books for various cities and he knows exactly what to do. He shows you the work he's done over the past few years: dozens of bids, all highly original and attractively presented. His agency, he says, offers nothing less than a bespoke, engaging bid proposal featuring all the right buzzwords, with a catchy motto that will make a big impression on the jury. The previous examples he shows you talk about 'open spaces', 'open minds' and 'frameworks', about 'empowerment' and 'urban labs'. And pretty much all the cities, you notice, are located 'at the heart of Europe' and are hubs of communication and cultural exchange. The consultant stands beside you, nods, and finally names a price for his art. Is this what people want? Ready-made originality? Doesn't this rob every word of its utopian potential? You ask yourself: doesn't a bid written in 2019 need to strike a completely different tone from one written five years ago, after all that has happened? Doesn't it need to respond directly to the cultural and political realities of our continent, to the rift running through Europe? Shouldn't it be about the genuine – and now so urgently necessary – coming together of people as equals?" (Hannover2025: 2). György Konrád, a former member of

The ECoC initiative is now out of kilter. It is a toxic spiral in which cities are under pressure to make their ECoC programme more and more extensive in terms of content, as well as space, time and, ultimately, finances. When the small candidate city of Zittau (just under 30,000 inhabitants) listed its Christmas market as a cultural event in the cultural calendar on its homepage, some people simply laughed at the city. This was out of line with the styled, hyper-innovative self-portrayals of the other candidate cities that are standard today.

The fact that cities with less money are *de facto* excluded from the competition contradicts the basic principles of the ECoC initiative. An example of this is the "SECOC" (Shaping European Capitals of Culture) workshop held in Wrocław in October 2018.[58] Participation in the seminar, which was again offered by representatives of the same circle of consultants, cost 1,550 Euros per participant (including accommodation and meals). Some candidate cities were unable or unwilling to pay this sum and therefore did not send any representatives to the seminar. However, not only did they miss out on practical information for their bid books, but also on crucial connections to the masterminds of the initiative. One of these masterminds was the consultant who finally coached Chemnitz, but who had previously appeared in the candidate city of Nuremberg, together with a member of the selection panel, to offer his services, thus completely blurring the boundaries between consultants and jurors (see Ritzer 2020).

Another of SECOC's workshop leaders is one of the officials responsible for ECoCs in the EC. The question is why an EC official is acting as an advisor at a seminar whose participation fee is so high that some candidate cities simply cannot afford to attend. Wouldn't it make much more sense for the EC, for its part, to offer qualification seminars where no city is excluded for financial reasons? The EC did launch a capacity building programme for this purpose in 2019, but this programme applies only to designated

---

the selection jury, describes the role of the consultants in a similar way: "What can you use to make a city fashionable? A city fashion designer is contracted for a lot of money. The master comes for a year, finds the local obsessives and their regular places. As a dreamy vociferous advertising expert, he guesses what will sell" (quoted in Mak 2020: 364, translated by the author).

58 See link (i).

ECoCs (not candidate cities); furthermore, it was thwarted by the Covid-19 pandemic and its continuation is unclear. However, it would be better for the competing cities if the EC offered support during the bidding process, i.e. before the title is awarded. But if ECoC experts are involved in the offering of support, it is important to ensure that they are impartial and do not hold any other positions in the ECoC carousel that may give them access to confidential information from individual bidding cities.

The designation of Chemnitz2025

The ECoC title was hastily awarded to Chemnitz, apparently because of pragmatic considerations. However, the title should not be awarded by a jury with potential conflicts of interest. No one knows how the accusations of bias were addressed. The ECoC Chemnitz2025 would also be on a much sounder footing if the accusations had been properly addressed and the findings communicated to the public. Now the bitter aftertaste of a possibly unfair competition procedure remains.

It cannot be denied that the selection panel allowed a political reason to emerge in the reasons for its selection in the two selection rounds. For example, the elimination of Dresden as a candidate city was justified as follows: "The need for the ECoC title and its legacy was not clearly articulated."[59] This raises the suspicion that the selection panel followed the mood of some media[60] during the bidding process, such as the German Cultural Council (Deutscher Kulturrat 2019: 14) and the *Süddeutsche Zeitung* (Heidtmann & Nimz 2019), which contrasted the two Saxon competing cities in a questionably simplistic way: poor Chemnitz, which has a problematic reputation, versus dazzling, privileged Dresden. This is reminiscent of the selection of the ECoC Marseille-Provence 2013, in whose final round the selection panel's decision for Marseille and against Bordeaux is said to have been "because Marseille deserves it". Afterwards it was widely stated that pity was truly not a legitimate selection criterion and that the jurors should instead strictly adhere to the established criteria.

---

59 The Expert Panel's report Pre-Selection Stage, Selection of the European Capital of Culture (ECoC) 2025 in Germany, p. 10: see link (j).
60 See the author's assessment in Prante 2020: 22.

Ulrich Fuchs, who conducted several workshops with the candidate cities on behalf of the *Cultural Foundation of the Federal States* (Kulturstiftung der Länder) as managing authority, comments as follows on the question of whether there was an "Eastern bonus": "But only if the bid from Chemnitz or Magdeburg is just as good as one from the three West German cities. Then it could also count that one says: Well, the last ECoC was in Western Germany with Essen and the Ruhr area, maybe the next one will be in Eastern Germany" (quoted in Reiche 2020). Whether Dresden really dropped out of the competition at an early stage against the background of this simplistic black-and-white portrayal in the media cannot be proven. In any event, many did not understand the selection panel's assessment.

No one wants to take the ECoC title away from Chemnitz. Most are certainly in favour of the approach in the bid book to become a "model for Europe" by seeking dialogue with right-wing elements through artistic encounters and new, innovative formats. The need for this became again alarmingly clear during the digital ECoC title ceremony on 28 October 2020, when citizens of Chemnitz even chanted right-wing slogans in live chat during the panel chairwoman's announcement speech. Apart from the conflict of interest, the city has delivered a good bid and will hopefully use its ECoC budget of 90 million Euros for many good innovations and formats in the cultural sector. But good governance structures also require sound management that stands up to public scrutiny in every respect.

Cultural policy consequences for the future

The transparency and compliance deficits that have emerged have damaged the ECoC initiative and disappointed many of the actors. Now it is important to see the crisis as an opportunity to work through the points of criticism and use them constructively for a reorientation of the initiative. "We need a renewal of content, more transparency, more quality control. And perhaps completely new formats" summarised Gottfried Wanger, former member of the selection panel, after the accusations about the designation of Chemnitz.[61]

---

61 Quoted in Ruf: 2020.

In general, the Covid-19 pandemic means that many aspects of the cultural sector need to be readjusted and renegotiated. At the political level, there must be more transparency and checks and balances through binding compliance rules.

In any event, the selection panel must be accessible to the public. The question is to what extent the programme as a whole needs to be scaled down in order to remain authentic. Here, one could consider limiting the budgets for the bidding process or even the opening ceremonies of the cities, which are sometimes excessive in terms of the money spent. The current catalogue of criteria is also far too extensive for smaller cities. It should be reduced, as should the scope of the bid books, which leads to redundancy and an unmanageable "projectitis".

It would be beneficial if there were support structures available for those cities that have at least reached the final round. It is true that the German candidate cities for ECoC 2025 unanimously emphasise – despite all the criticism of the selection procedure – that they have grown through the competition. They have learnt a lot in terms of international cooperation, the acquisition of funding, and generally looking beyond their own borders. No other judgement would be expected, though, because after all, considerable municipal funds were invested in the bid and one would like to avoid this being seen by the citizens as a bad investment without results. The city of Nuremberg alone spent over 6.5 million Euros on its bid.[62] The value of participating in the ECoC competition can only be assessed in a few years. But especially due to the consequences of the Covid-19 pandemic, and the associated cuts in the cultural sector, there is a danger that the opportunities gained will be lost again. In Italy, each candidate city received 1 million Euros for the implementation of its Plan B. In Germany, the federal government does not offer such financial support. The EU should support such agreements, or could itself also financially support the finalists.

The "consultant-gate" revealed that it is good for the participating cities to trust in their local potential. By reflecting on the capacity in their surroundings, the local self-confidence of the city and the region is promoted – and a lot of CO_2 is saved.

---

62 See link (k).

The meetings of the so-called ECoC Family, i.e. those responsible for past, present and future ECoCs, should no longer take place behind closed doors. Process observers, for example, from academia, should be allowed to attend such meetings. Likewise, process observers should also be allowed to participate in the qualification seminars of the "managing authority".

The German candidate cities would have liked more specific feedback on their years of preparatory work and were dissatisfied with the lack of detail in the jury reports. More in-depth explanations, for example, in conversational form, would improve the imbalance between the complex bidding process and the very brief reasons for the judgement.

No goal has been idealistic enough for the ECoCs of recent years. "What can Europe learn from us, and what can we learn from Europe? How do we want to live together in the future? What can we do better to implement the ideals of Europe?" are the standard questions repeatedly addressed at the relevant ECoC panels. Fundamental values such as identity, solidarity and humanity must be strengthened through participation in the ECoC programme. Through artistic interventions, reflections on the current, comprehensive tasks in Europe such as climate change, migration, social division, the rural exodus and many more must be stimulated; "no goal was too ambitious" (Mak 2020: 365). Often this is formulated in such general terms that no one can disagree. That is all well and good. But in taking such a moral high ground, the ECoC programme itself must also act with integrity.

Cronyism and lack of transparency contradict the noble goals of the initiative. The emotionally charged claims of a Europe united in diversity were misused as a smokescreen and concealed questionable business relationships existing in the background. By making the ECoC objectives so consensual, critics could be silenced. Thus, the credibility of the laudable programme is in danger. But art and culture in particular should offer an unquestionable morality in times of social reorganisation.

The ECoC initiative comprises far more than the jury caught up in conflicts of interest, "a roving group that had been informed and courted by the rival cities and had developed very specific, clearly discernible preferences and sensitivities over the years" and "an equally international itinerant group of experts and copywriters who knew exactly what the jury wanted to see and hear and advised the candidate cities in return for generous fees" (Mak 2020: 365). The ECoC initiative above all comprises many committed citizens in the participating cities who put their heart and soul into a fruitful European peace and culture project. In order for the cultural sector to continue to be a moral authority that creates touching moments in our thoroughly rationalised world, and that, contrary to all moves towards isolation, re-nationalisation and homogenisation, focuses on what unites us and on our peaceful diversity in Europe, the ECoC initiative needs a solid and unassailable foundation.

Kristina Marion Jacobsen
Berlin, summer 2021

Bibliography

Deutscher Kulturrat (2019): Wer wird Kulturhauptstadt Europas 2025? Die Redaktion benennt ihre Favoriten 14, in: Politik und Kultur 12/2019–01/2020. Available online at: https://www.kulturrat.de/wp-content/uploads/2019/11/puk1219-0120.pdf

Hannover2025: Agora of Europe. Bid book (2020) Available online at: https://www.khh25.de/agoraofeurope

Hanssen, Frederik (2021): Streit um die Kulturhauptstadt. Was passiert hinter den Kulissen?, in: Tagesspiegel, 12 January 2021. Available online at: https://www.tagesspiegel.de/kultur/streit-um-die-kulturhauptstadt-was-passiert-hinter-den-kulissen/26791620.html

Heidtmann, Jan & Nimz, Ulrike (2019): Sachsen: Dresden oder Chemnitz?, in: Süddeutsche Zeitung, 6 December 2019. Available online at: https://www.sueddeutsche.de/kultur/rechtsradikale-kulturhauptstadt-dresden-chemnitz-1.4709028?reduced=true

Jacobsen, Kristina Marion (2009): Die drei Europäischen Kulturhauptstädte in Deutschland im Kontext der Entwicklung europäischer Kulturpolitik, Saarbrücken

Jacobsen, Kristina (2020): Eight Cities, One Goal: The Application Process of German Cities as ECoC 2025, in: Matiu, Ovidiu & Farrugia, Glen (eds.): Cultural Resilience: Physical Artifacts, Intangible Attributes, Natural Risks – Proceedings of the Thirteenth Interdisciplinary Conference of the University Network of the European Capitals of Culture, Matera, Italy, 28–29 November 2019, UNEECC FORUM VOLUME 12. Lucian Blaga University of Sibiu Press, Sibiu, pp. 141–148. Available online at: https://uneecc.org/conference/proceedings/

Kauffmann, Bernd (2020): Hannover hatte keine Chance auf die Kulturhauptstadt, Interview, in: Neue Presse, 8 December 2020. Available online at: https://www.neuepresse.de/Nachrichten/Kultur/Uebersicht/Kulturhauptstadt-Hannover-Interview-Bernd-Kauffmann

Mak, Geert (2020): Große Erwartungen: Auf den Spuren des europäischen Traums, München, pp. 364–369

Prante, Martina (2020): Wettbewerb verloren: Blick nach vorne!, Interview with Kristina Jacobsen, in: Hildesheimer Zeitung, 5 November 2020. Available online at: https://www.hildesheimer-allgemeine.de/meldung/nachgefragt-war-die-hildesheimer-bewerbung-vielleicht-zu-gut.html

Ritzer, Uwe (2021): Das Amigo-System beenden, in: Süddeutsche Zeitung, 2 January 2021, p. 4

Ritzer, Uwe (2020): Das Geschäft hinter dem Titel, in: Süddeutsche Zeitung, 3 December 2020. Available online at: https://www.sueddeutsche.de/kultur/kulturhauptstadt-kulturhauptstadt-2025-chemnitz-berater-1.5136643?reduced=true

Reiche, Matthias (2020): Fünf deutsche Städte hoffen auf den Titel, Interview mit Ulrich Fuchs, 28 October 2020. Available online at: https://www.tagesschau.de/ausland/kulturhauptstadt-europas-101.html

Ruf, Birgit (2020): Kulturhauptstadt-Verfahren: Intransparentes Millionenspiel, in: Nordbayern, 2 December 2020. Available online at: https://www.nordbayern.de/kultur/kulturhauptstadt-verfahren-intransparentes-millionenspiel-1.10647382?tabParam=comments

Links

(a) https://www.hannover.de/Kulturhauptstadt-Hannover/Aktuelles/HIER-JETZT-ALLE-f%C3%BCr-Europa

(b) https://www.magdeburg2030.de/veranstaltungen/detail/news/ecoc-konferenz-under-construction-bewerberstaedte-zur-kulturhauptstadt-2025-treffen-sich-in-magdeburg/

(c) https://chemnitz2025.de/bidbook/

(d) https://www.kmk.org/aktuelles/artikelansicht/chemnitz-wird-kulturhauptstadt-europas-im-jahr-2025-kultur-ministerkonferenz-folgt-expertenvotum-d.html and https://www.kmk.org/fileadmin/veroeffentlichungen_beschluesse/2021/2021_01_11-Kulturhauptstadt-Europa.pdf

(e) https://www.tagesschau.de/multimedia/sendung/tt-8011.html

(f) https://eur-lex.europa.eu/legal-content/EN/TXT/?uri=celex%3A32014D0445

(g) https://ec.europa.eu/culture/policies/culture-cities-and-regions/
european-capitals-culture

(h) https://en.wikipedia.org/wiki/Perverse_incentive#The_original_
cobra_effect

(i) https://www.secoc.eu/

(j) https://ec.europa.eu/culture/document/
pre-selection-report-european-capital-culture-2025-germany

(k) https://www.br.de/nachrichten/kultur/auch-ohne-titel-nuernberg-
fuehrt-kulturhauptstadt-projekte-weiter,SOY1is7

All links listed here were last accessed in June 2021.

Studien zur Kulturpolitik
Cultural Policy

Herausgegeben von / Edited by Prof. Dr. Wolfgang Schneider

www.peterlang.com